Historical and Current Diversity Patterns of Mediterranean Marine Species

Historical and Current Diversity Patterns of Mediterranean Marine Species

Editor

Sabrina Lo Brutto

MDPI • Basel • Beijing • Wuhan • Barcelona • Belgrade • Manchester • Tokyo • Cluj • Tianjin

Editor
Sabrina Lo Brutto
Department of Biological,
Chemical, and Pharmaceutical
Science and Technology
(STeBiCeF)
University of Palermo
Palermo
Italy

Editorial Office
MDPI
St. Alban-Anlage 66
4052 Basel, Switzerland

This is a reprint of articles from the Special Issue published online in the open access journal *Diversity* (ISSN 1424-2818) (available at: www.mdpi.com/journal/diversity/special_issues/marine_mediterranean_species).

For citation purposes, cite each article independently as indicated on the article page online and as indicated below:

LastName, A.A.; LastName, B.B.; LastName, C.C. Article Title. *Journal Name* **Year**, *Volume Number*, Page Range.

ISBN 978-3-0365-1205-1 (Hbk)
ISBN 978-3-0365-1204-4 (PDF)

Contents

About the Editor . . . . **vii**

Preface to "Historical and Current Diversity Patterns of Mediterranean Marine Species" . . . **ix**

Sabrina Lo Brutto
Historical and Current Diversity Patterns of Mediterranean Marine Species
Reprinted from: *Diversity* **2021**, *13*, 156, doi:10.3390/d13040156 . . . . **1**

Sabrina Lo Brutto, Andrea Calascibetta, Gianni Pavan and Gaspare Buffa
Cetacean Strandings and Museum Collections: A Focus on Sicily Island Crossroads for
Mediterranean Species
Reprinted from: *Diversity* **2021**, *13*, 104, doi:10.3390/d13030104 . . . . **5**

**Maria Flavia Gravina, Andrea Bonifazi, Michela Del Pasqua, Jacopo Giampaoletti, Marco
Lezzi, Daniele Ventura and Adriana Giangrande**
Perception of Changes in Marine Benthic Habitats: The Relevance of Taxonomic and Ecological
Memory
Reprinted from: *Diversity* **2020**, *12*, 480, doi:10.3390/d12120480 . . . . **21**

**Barbara Mikac, Margherita Licciano, Andrej Jaklin, Ljiljana Iveša, Adriana Giangrande and
Luigi Musco**
Diversity and Distribution Patterns of Hard Bottom Polychaete Assemblages in the North
Adriatic Sea (Mediterranean)
Reprinted from: *Diversity* **2020**, *12*, 408, doi:10.3390/d12100408 . . . . **37**

**Francesco Luigi Leonetti, Emilio Sperone, Andrea Travaglini, Angelo R. Mojetta, Marco
Signore, Peter N. Psomadakis, Thaya M. Dinkel and Massimiliano Bottaro**
Filling the Gap and Improving Conservation: How IUCN Red Lists and Historical Scientific
Data Can Shed More Light on Threatened Sharks in the Italian Seas
Reprinted from: *Diversity* **2020**, *12*, 389, doi:10.3390/d12100389 . . . . **57**

**Tatiana Fioravanti, Andrea Splendiani, Tommaso Righi, Nicola Maio, Sabrina Lo Brutto,
Antonio Petrella and Vincenzo Caputo Barucchi**
A Mediterranean Monk Seal Pup on the Apulian Coast (Southern Italy): Sign of an Ongoing
Recolonisation?
Reprinted from: *Diversity* **2020**, *12*, 258, doi:10.3390/d12060258 . . . . **73**

**Tommaso Righi, Andrea Splendiani, Tatiana Fioravanti, Elia Casoni, Giorgia Gioacchini,
Oliana Carnevali and Vincenzo Caputo Barucchi**
Loss of Mitochondrial Genetic Diversity in Overexploited Mediterranean Swordfish (*Xiphias
gladius*, 1759) Population
Reprinted from: *Diversity* **2020**, *12*, 170, doi:10.3390/d12050170 . . . . **81**

About the Editor

Sabrina Lo Brutto

Sabrina Lo Brutto received her PhD in Animal Biology at the University of Palermo (Italy). After a Post-doctoral fellowship on Fish Population Genetics, she became a Researcher, and is now Professor in Zoology and Director of the Museum of Zoology Doderlein. Her past and recent research areas of interest have been Phylogeography and Population Genetics of European invertebrates and vertebrates species. She worked mostly on marine Mediterranean taxa, including invasive alien species. Her recent research focuses on the Taxonomy, Ecology and Phylogeography of the crustacean amphipods. Her research ranges from traditional morpho-taxonomic approaches to molecular ecology studies. Her interest is also addressed to historical animal collections and Natural History Museums. She is author of many papers published in international peer-reviewed journals on Marine Biology, Zoology and Museology and she is a reviewer for the most relevant journals in such fields.

Preface to "Historical and Current Diversity Patterns of Mediterranean Marine Species"

The Mediterranean Sea is a semi-enclosed basin that experienced different natural and anthropogenic phenomena, which produced biological diversity changes over time. The Mediterranean Sea has been through dramatic changes in its biota over the last 6 million years, and more rapidly in the past century. All the events left a footprint on the gene pool of marine species, on their morpho-physiological traits, and on the loss or expansion of the geographical range extent. Nowadays, the Mediterranean is changing its physical and ecological features. The changes in its environmental conditions are followed by changes in species composition, of which most of the heritage has been stored and preserved in historical museum collections. In this Book, the biodiversity of the Mediterranean Sea is described at a synchronic and at a diachronic level, as the biodiversity of an area needs to be described on a spatial and temporal scale. The Book focuses on data relating to the past two centuries for which museum collections can provide overlooked information. Available information exists for the major marine taxa, knowledge of which would greatly benefit from materials and data collected in the past. The scientific community has a relevant role in the present day. It must not leave sporadic and fragmented efforts and must instead move towards a coordinated framework. We all know how much biodiversity is relevant for human life, and we have to be conscious of how much information we still have as unpublished data. Here, seven papers depict the Mediterranean Biodiversity from different points of view; they cover different topics and different taxa (seagrasses, macroalgae, sponges, polychaetes, bivalves, sharks, fishes, mammals). The Book can be also a tool for educational purposes, as several case-studies reveal unexpected patterns. My thanks go to the publishers and to all the authors who sent their contribution to the Special Issue—Historical and Current Diversity Patterns of Mediterranean Marine Species—printed here. I greatly appreciated their work and I am sure you will do as well.

Sabrina Lo Brutto
Editor

 *diversity*

Editorial

Historical and Current Diversity Patterns of Mediterranean Marine Species

Sabrina Lo Brutto [1,2]

1   Department of Biological, Chemical, and Pharmaceutical Science and Technology (STEBICEF),
    University of Palermo, 90123 Palermo, Italy; sabrina.lobrutto@unipa.it
2   Museum of Zoology "P. Doderlein", SIMUA, University of Palermo, 90123 Palermo, Italy

check for
updates

Citation: Lo Brutto, S. Historical and Current Diversity Patterns of Mediterranean Marine Species. *Diversity* **2021**, *13*, 156. https://doi.org/10.3390/d13040156

Received: 1 April 2021
Accepted: 1 April 2021
Published: 6 April 2021

The Mediterranean is a sea which, despite its peculiar geomorphological history and ecological–oceanographic features, still receives less attention than it ought to. The Mediterranean is a semi-enclosed basin where different natural and anthropogenic phenomena have caused changes at the community or intra-species level over time. The basin went through dramatic changes in its biota during the last six million years and more quickly in the past century [1]. The contemporary physical factors, such as hydrodynamic patterns, temperature and salinity, further affect species distribution, interacting with biological factors such as bioinvasions [2].

All the events and all the processes have left and leave a mark in the gene pool of marine species [3], their morpho-physiological traits [4], and the extent of the loss or expansion of their geographic range [5].

Currently, the Mediterranean Sea is changing its physical and ecological features, whose trends can be extrapolated by comparing historical collections and present-day observations. Natural history museums have a fundamental role in this research field as they preserve historical biodiversity and reference material of a region [6]. Notwithstanding, they have been overlooked by academic researchers for a long time.

The Mediterranean has been described as a miniature ocean [7] for its species richness and the overall response to the diverse pressures affecting its biota. Accordingly, the basin can be considered a natural laboratory, and the case studies from this marine realm can be symbolic of specific topics [8].

Recently, a growing number of studies has focused on themes such as alien or vagrant species spread [2,9] or warming climate forecasts [10,11]. Scarce literature deals with patterns obtained from long-term datasets or museum collections which can provide information on a large temporal scale, while few articles report relevant data on an extensive variety of species, in some cases focusing on conservation purposes [12,13]. Consequently, more research is needed to fill these gaps and provide additional information.

In this issue, the biodiversity in the Mediterranean Sea has been described at a synchronic and a diachronic level, highlighting the past two centuries for which museum collections can provide overlooked information. Historical records are preserved for the major marine taxa, knowledge of which would greatly benefit from employing specimens and data collected in the past. All of the articles review the current status of the marine diversity of species belonging to several taxonomic groups (seagrasses, macroalgae, sponges, polychaetes, bivalves, sharks, fishes, mammals) and explore the ecological and conservation implications of some of the most threatened ones.

A study examined the extent of the cetacean strandings in Italy [14]. The authors estimated the number of marine cetacean strandings by means of a long-term dataset, covering a thirty-year period. The pattern described reflects the knowledge on the distribution of common and rare cetacean species and raises some questions about the necessity to organize the recovery of carcasses in some regions, to not lose samples and data.

The Mediterranean Sea hosts the Mediterranean monk seal (*Monachus monachus*), one of the most endangered marine mammals in the world. It is a charismatic species very close to extinction. However, an increase in sporadic sightings has been recorded in recent years. The paper by Fioravanti et al. [15] reports the results of the genetic characterization of a monk seal pup stranded on the southern Italian coast, underlining the need to intensify conservation activities for this species as it could be much more widespread than previously thought.

Sharks are also one of the most threatened marine animal groups worldwide, and the Mediterranean Sea is considered an extinction hotspot for such species. Historical data have provided important information on how chondrichthyan populations have changed over time. A study included in this Special Issue [16] focuses on some selected species, for which a bibliographic search was conducted on the literature from the 19th century to the first half of the 20th century. The results showed that all the sharks were considered common until the beginning of the 20th century but have declined for the past 70 years. The authors attributed the strong decline to overexploitation, bycatch, habitat loss, depletion of prey items, and environmental pollution.

The above historical pattern drives us to deal with the issue of shifting baseline syndrome (SBS) [17], a behavior of new-generation scientists with a lack of perception of the past ecological conditions and changing ecosystems. In other words, most young people are not conscious of how abundant some species were and how much they have declined today. The paper by Gravina et al. [17] highlights that having a reliable ecological reference baseline is pivotal to understanding the current status of marine biodiversity. Ecological awareness of our perception of environmental changes could be better described based on historical data. Combining historical data with contemporary biomonitoring is required for conservation strategies. The authors advocate for the crucial role of taxonomy as a study of life diversity and the informative value of museum collections as memories of past ecosystem conditions. The paper [17] focuses on six Mediterranean benthic habitats to track biological and structural changes that have occurred in the last few decades.

Herbaria and zoological collections are certainly fundamental for taxonomic studies, and they are also invaluable, though currently underestimated, resources for understanding ecological and evolutionary responses of species to environmental changes. In particular, macroalgae herbarium collections, which exist in some European herbaria, can be successfully used as real "witnesses" to biodiversity changes [18].

Investigations on the temporal genetic variation within a species are also relevant in stock assessment studies. The paper by Righi et al. [19] describes the complex situation of one of the most exploited fish species, the Mediterranean swordfish, of which abundance has drastically decreased. The possible relationship between fishery activities and the loss of genetic diversity in the Mediterranean fish populations is a further crucial point.

The diversity of hard bottom fauna is also largely underestimated and needs regular updating in order to detect and monitor changes in benthic communities. For this reason, Mikac et al. [20] contributed to updating the information on polychaete diversity and depicted a pattern of spatial variation in relation to changes in algal coverage at increasing depth.

In conclusion, this Special Issue filled some gaps, though the Mediterranean Sea still remains an unexplored basin regarding some taxa and some areas, as geopolitics influence the collection of data, and inadequate funds limit the survey of peculiar habitats such as the deep sea. The Mediterranean Sea is currently experiencing a decline in the abundance of several key species, as a consequence of anthropogenic pressures (increase in human population, habitat modification and loss, pollution, coastal urbanization, overexploitation, introduction of non-indigenous species, and climate change), and the scientific community has a relevant role in the present day. It should not have to rely on sporadic and fragmented efforts towards an uncoordinated framework. We all know how much biodiversity is relevant for human life, and how much information still needs to be discovered and organized.

Funding: This study was supported by the University of Palermo.

Conflicts of Interest: The author declares no conflict of interest.

References

1. Bianchi, C.N.; Morri, C.; Chiantore, M.; Montefalcone, M.; Parravicini, V.; Rovere, A. Mediterranean Sea biodiversity between the legacy from the past and a future of change. In *Life in the Mediterranean Sea: A Look at Habitat Changes*; Stambler, N., Ed.; Nova Science Publishers Inc.: New York, NY, USA, 2012; Volume 1, p. 55.
2. Servello, G.; Andaloro, F.; Azzurro, E.; Castriota, L.; Catra, M.; Chiarore, A.; Crocetta, F.; D'Alessandro, M.; Denitto, F.; Froglia, C.; et al. Marine alien species in Italy: A contribution to the implementation of Descriptor D2 of the marine strategy framework directive. *Mediterr. Mar. Sci.* **2019**, *20*, 1. [CrossRef]
3. Maggio, T.; Lo Brutto, S.; Cannas, R.; Deiana, A.M.; Arculeo, M. Environmental features of deep-sea habitats linked to the genetic population structure of a crustacean species in the Mediterranean Sea. *Mar. Ecol.* **2009**, *30*, 354–365. [CrossRef]
4. Mejri, R.; Lo Brutto, S.; Hassine, N.; Arculeo, M.; Hassine, O.K.B. Overlapping patterns of morphometric and genetic differentiation in the Mediterranean goby *Pomatoschistus tortonesei* Miller, 1968 (Perciformes, Gobiidae) in Tunisian lagoons. *Zoology* **2012**, *115*, 239–244. [CrossRef] [PubMed]
5. Lo Brutto, S.; Iaciofano, D.; Guerra García, J.M.; Lubinevsky, H.; Galil, B.S. Desalination effluents and the establishment of the non-indigenous skeleton shrimp *Paracaprella pusilla* Mayer, 1890 in the south-eastern Mediterranean. *BioInvasions Rec.* **2019**, *8*, 661–669. [CrossRef]
6. Lo Brutto, S. The case of a rudderfish highlights the role of natural history museums as sentinels of bio-invasions. *Zootaxa* **2017**, *3*, 382–386. [CrossRef] [PubMed]
7. Lejeusne, C.; Chevaldonné, P.; Pergent-Martini, C.; Boudouresque, C.F.; Pérez, T. Climate change effects on a miniature ocean: The highly diverse, highly impacted Mediterranean Sea. *Trends Ecol. Evol.* **2010**, *25*, 250–260. [CrossRef] [PubMed]
8. Lo Brutto, S.; Iaciofano, D.; Lo Turco, V.; Potortì, A.G.; Rando, R.; Arizza, V.; Di Stefano, V. First assessment of plasticizers in marine coastal litter-feeder fauna in the Mediterranean Sea. *Toxics* **2021**, *9*, 31. [CrossRef] [PubMed]
9. Mannino, A.M.; Balistreri, P.; Iaciofano, D.; Galil, B.S.; Lo Brutto, S. An additional record of *Kyphosus vaigiensis* (Quoy & Gaimard, 1825) (Osteichthyes, Kyphosidae) from Sicily clarifies the confused situation of the Mediterranean kyphosids. *Zootaxa* **2015**, *3963*, 45–54. [PubMed]
10. Azzurro, E.; Sbragaglia, V.; Cerri, J.; Bariche, M.; Bolognini, L.; Ben Souissi, J.; Busoni, G.; Coco, S.; Chryssanthi, A.; Fanelli, E.; et al. Climate change, biological invasions, and the shifting distribution of Mediterranean fishes: A large-scale survey based on local ecological knowledge. *Glob. Chang. Biol.* **2019**, *25*, 2779–2792. [CrossRef] [PubMed]
11. Sarà, G.; Milanese, M.; Prusina, I.; Sarà, A.; Angel, D.L.; Glamuzina, B.; Nitzan, T.; Freeman, S.; Rinaldi, A.; Palmeri, V.; et al. The impact of climate change on Mediterranean intertidal communities: Losses in coastal ecosystem integrity and services. *Reg. Environ. Change* **2014**, *14*, 5–17. [CrossRef]
12. Landi, M.; Dimech, M.; Arculeo, M.; Biondo, G.; Martins, R.; Carneiro, M.; Carvalho, G.R.; Lo Brutto, S.; Costa, F.O. DNA barcoding for species assignment: The case of Mediterranean marine fishes. *PLoS ONE* **2014**, *9*, e106135. [CrossRef] [PubMed]
13. Cariani, A.; Messinetti, S.; Ferrari, A.; Arculeo, M.; Bonello, J.J.; Bonnici, L.; Cannas, R.; Carbonara, P.; Cau, A.; Charilaou, C.; et al. Improving the conservation of Mediterranean chondrichthyans: The Elasmomed DNA barcode reference library. *PLoS ONE* **2017**, *12*, e0170244. [CrossRef] [PubMed]
14. Lo Brutto, S.; Calascibetta, A.; Pavan, G.; Buffa, G. Cetacean strandings and museum collections: A focus on Sicily Island crossroads for Mediterranean species. *Diversity* **2021**, *13*, 104. [CrossRef]
15. Fioravanti, T.; Splendiani, A.; Righi, T.; Maio, N.; Lo Brutto, S.; Petrella, A.; Caputo Barucchi, V. A Mediterranean monk seal pup on the Apulian Coast (Southern Italy): Sign of an ongoing recolonisation? *Diversity* **2020**, *12*, 258. [CrossRef]
16. Leonetti, F.L.; Sperone, E.; Travaglini, A.; Mojetta, A.R.; Signore, M.; Psomadakis, P.N.; Dinkel, T.M.; Bottaro, M. Filling the gap and improving conservation: How IUCN red lists and historical scientific data can shed more light on threatened sharks in the Italian Seas. *Diversity* **2020**, *12*, 389. [CrossRef]
17. Gravina, M.F.; Bonifazi, A.; Del Pasqua, M.; Giampaoletti, J.; Lezzi, M.; Ventura, D.; Giangrande, A. Perception of changes in marine benthic habitats: The relevance of taxonomic and ecological memory. *Diversity* **2020**, *12*, 480. [CrossRef]
18. Mannino, A.M.; Armeli Minicante, S.; Rodríguez-Prieto, C. Phycological Herbaria as a useful tool to monitor long-term changes of macroalgae diversity: Some case studies from the Mediterranean Sea. *Diversity* **2020**, *12*, 309. [CrossRef]
19. Righi, T.; Splendiani, A.; Fioravanti, T.; Casoni, E.; Gioacchini, G.; Carnevali, O.; Caputo Barucchi, V. Loss of mitochondrial genetic diversity in overexploited Mediterranean Swordfish (*Xiphias gladius*, 1759) population. *Diversity* **2020**, *12*, 170. [CrossRef]
20. Mikac, B.; Licciano, M.; Jaklin, A.; Iveša, L.; Giangrande, A.; Musco, L. Diversity and distribution patterns of hard bottom polychaete assemblages in the North Adriatic Sea (Mediterranean). *Diversity* **2020**, *12*, 408. [CrossRef]

Review

Cetacean Strandings and Museum Collections: A Focus on Sicily Island Crossroads for Mediterranean Species

Sabrina Lo Brutto [1,2,3,*], Andrea Calascibetta [1,3], Gianni Pavan [4] and Gaspare Buffa [5]

[1] Department of Biological, Chemical, and Pharmaceutical Science and Technology (STEBICEF), University of Palermo, 90123 Palermo, Italy; andreacalascibetta91@gmail.com
[2] Centro Interuniversitario di Ricerca sui Cetacei (CIRCE), Unit of the University of Palermo, 90123 Palermo, Italy
[3] Museum of Zoology "P. Doderlein", SIMUA, University of Palermo, 90123 Palermo, Italy
[4] Department of Earth and Environment Sciences, University of Pavia, 27100 Pavia, Italy; gianni.pavan@unipv.it
[5] National Research Council-Institute of Anthropic Impact and Sustainability in Marine Environment-(CNR-IAS), Section of Capo Granitola, Campobello di Mazara (TP), 91021 Sicily, Italy; gaspare.buffa@ias.cnr.it
* Correspondence: sabrina.lobrutto@unipa.it

Abstract: The study examined the extent of the cetacean strandings in Italy, with a particular focus on Sicily Island. The paper aimed to contribute to the description of a pattern that contemplates the "regular and rare" cetacean species passage along the Sicilian coast. The estimate of marine cetacean strandings was extrapolated from the National Strandings Data Bank (BDS—Banca Dati Spiaggiamenti) and evaluated according to a subdivision in three coastal subregions: the Tyrrhenian sub-basin (northern Sicilian coast), the Ionian sub-basin (eastern Sicilian coast), and the Channel of Sicily (southern Sicilian coast). Along the Italian coast, more than 4880 stranding events have been counted in the period 1990–2019. Most of these were recorded in five Italian regions: Apulia, Sicily, Sardinia, Tuscany, and Calabria. Approximately 15% of the recorded strandings in Italy occurred on the Sicilian coast. In Sicily Island, 725 stranded cetaceans were recorded in 709 stranding events, resulting in approximately 20 carcasses every year; the total number of specimens identified to species level was 539. The distribution along the Sicilian coast was the following: 312 recorded in the Tyrrhenian sub-basin, 193 in the Ionian sub-basin, and 220 in the Channel of Sicily. *Stenella coeruleoalba* was the species that can be considered as the stable record along the time-lapse investigated, and some rare species have been recorded as well. The role of Sicily Island as a *sentinel territory* of the cetacean distribution for the central Mediterranean Sea and as a region receiving a marine resource suitable for the scientific research and cetological museum collections is discussed herein.

Keywords: marine mammals; cetacean strandings; natural history museums; zoological collections; Mediterranean biodiversity

Citation: Lo Brutto, S.; Calascibetta, A.; Pavan, G.; Buffa, G. Cetacean Strandings and Museum Collections: A Focus on Sicily Island Crossroads for Mediterranean Species. *Diversity* **2021**, *13*, 104. https://doi.org/10.3390/d13030104

Academic Editor: Bert W. Hoeksema

Received: 31 December 2020
Accepted: 21 February 2021
Published: 26 February 2021

Publisher's Note: MDPI stays neutral with regard to jurisdictional claims in published maps and institutional affiliations.

1. Introduction

Natural history museums play an important role in zoological research and in the dissemination of scientific results to society, as well as in the enhancement of historical collections that mirror the past and current biodiversity of a region [1–3]. This is particularly valuable for cetacean collections [4,5] that preserve specimens which have been caught and specimens which have been stranded or died as a result of accidental captures in fishing gear (bycatch and entanglement) [6–8]. Some marine cetaceans in Italian museums, collected after stranding events, are remarkable for the Mediterranean Sea. An example is a specimen of the northern right whale, *Eubalaena glacialis*, a rare species for the basin, which stranded in Taranto in 1877 [9], and is presently exposed at the Zoological Museum of the University of Naples Federico II [10]. Two specimens of the Indo-Pacific rough-toothed dolphin, *Steno bredanensis*, recently considered a common species in the eastern

Mediterranean Sea [11], were stranded on the Sicilian coast in 2002 and now exhibited at the Civic Museum of Natural History of Comiso [12]. In addition, an exceptional skeleton of the dwarf sperm whale, *Kogia sima*, stranded near Foce Chiarore, Capalbio (Grosseto, Tuscany) in 1988 [13], which represents the first record for the Mediterranean, is shown at the Natural History Museum of the Accademia dei Fisiocritici of Siena.

Generally, cetaceans occupy a prominent position in museum exhibition halls and represent a great attraction to the public. They are considered *totem animals* [14] owing to their high emotional impact and because they function as precious documentary material useful for research and science dissemination. The abundant cetological heritage of Italian museums has been collected thanks to the work of scholars and taxidermists since the early 18th century [15] (see Supplement Document F1). The most important cetological collection in Italy in terms of the number of species and taxonomic diversity is now exhibited at the Natural History Museum of Calci (Pisa, Italy) where three species from Sicily Island are stored: a complete skeleton of a mounted sperm whale, *Physeter macrocephalus* (site collection Isola Grande—Marsala, Trapani, 1892); a skull and complete mounted skeleton of a Risso's dolphin, *Grampus griseus* (site collection Palermo 1881); and two skulls of the false killer whale, *Pseudorca crassidens* (date still unknown, previous 1900) [16].

The numerous specimens originating from Sicily and preserved in the Italian zoological collections, as well as the increasing scientific interest and public sensibility towards marine life by society led us to be aware of the extent of strandings along the coast of Sicily Island, the southernmost Italian region, in the view of offering a supporting document to whom have to manage strandings and plan a systematic collection of museum materials.

The present paper shows an assessment of the cetacean strandings that occurred in Sicily in the period 1990–2019 (Figure 1).

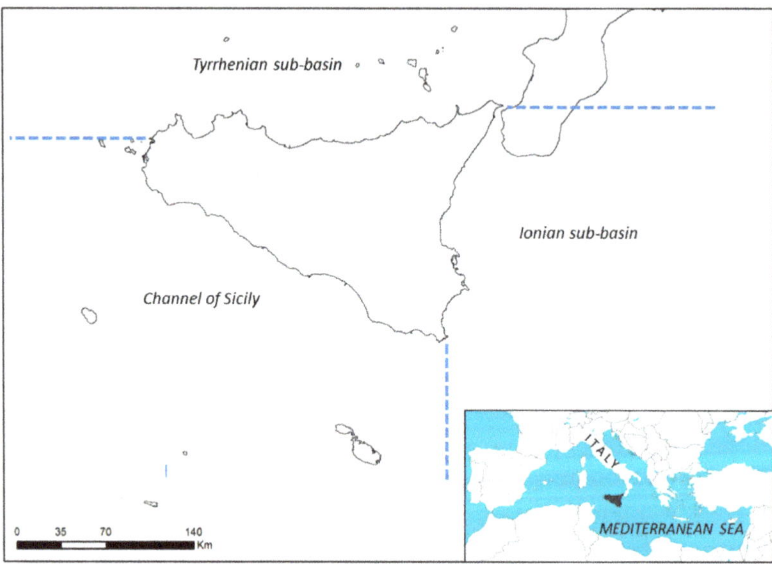

Figure 1. Map showing the partitioning of the Sicilian coast for the assessment of cetacean strandings: the Tyrrhenian sub-basin (northern Sicilian coast), the Ionian sub-basin (eastern Sicilian coast), and the Channel of Sicily (southern Sicilian coast). The insert indicates the position of Sicily Island in the Mediterranean Sea.

The data have been extracted from the National Strandings Data Bank (BDS—Banca Dati Spiaggiamenti) [17,18], and have been discussed in relation to the total stranding events along the Italian peninsula. The assessment was based on a coastal subdivision outlined in previous literature [19]; the Sicilian coast was partitioned into three sectors: the

Tyrrhenian sub-basin (the northern Sicilian coast), the Ionian sub-basin (the eastern Sicilian coast), and the Channel of Sicily (the southern Sicilian coast) (Figure 1).

Though the present paper reports the first description of spatial and temporal stranding records on the Italian coast, it does not deal with the impacts of humans on cetaceans or reasons for their mortality.

2. The National Strandings Data Bank (BDS–Banca Dati Spiaggiamenti)

The Italian research community benefits from organized information on the marine mammal strandings thanks to the National Strandings Data Bank [17,18]. The first Italian Strandings Network was created in 1986 at the Natural History Museum of Milan along with the Centro Studi Cetacei (CSC), a voluntary association of cetacean experts belonging to the Italian Society for Nature Sciences. Twenty years later, in 2006, the National Strandings Data Bank (BDS—Banca Dati Spiaggiamenti) was created and made available online by the University of Pavia and the Natural History Museum of Milan on behalf of the Italian Ministry of the Environment. The online data bank collects and validates strandings data to be made available to Governmental and Research Institutions as well as to the general public.

The BDS (http://mammiferimarini.unipv.it (accessed on 31 December 2020)) holds the records published by the Centro Studi Cetacei in the years 1986–2006, and since 2006, it has been updated in real-time with data sent by the Italian Strandings Network, which is managed by the Italian Ministry of the Environment and by the Ministry of Health. Currently, the initial reports of stranded animals are collected by the Coast Guard to be verified, validated, and transmitted to the competent territorial bodies and to the BDS. Any notice of stranded animals from citizens should be addressed to the Coast Guard.

The BDS also incorporates some historical data collected from a previous Cetacean Project, which was operative since 1975 and then merged into the CSC project. Data from Tuscany have, since 2007, been reported to the BDS directly by the regional network for the recovery of animals stranded along the Tuscan coast (Tuscan Observatory for Biodiversity—OTB) (L.R. n.30/2015, art. 11; Official Bulletin of the Tuscany Region n. 14 of 25 March 2015).

The scientific committee of the CSC also produced annual reports published by the Natural History Museum of Milan (Atti Soc. Ital. Sci. nat. Museo civ. Stor. Nat. Milano from I of 1986 to XXI of 2012; see Supplement Document F2).

In the period 1986 to 2019, 5571 stranding events were recorded in the BDS, totaling 6690 stranded animals, 4832 of which belong to 14 species and the rest were not identified. The records also include dead animals that were found entangled in fishing nets or on beaches and live animals that were caught in nets and then released.

The project National Stranding Data Bank (BDS) is within the frame of the activities recommended by ACCOBAMS (Agreement on the Conservation of Cetaceans of the Black Sea, Mediterranean and Contiguous Atlantic Area) and by the European Marine Strategy to monitor cetacean populations, the impacts of human activities, and the quality of the marine environment. The BDS is managed by the University of Pavia (CIBRA/Department of Earth and Environment Sciences) and by the Museum of Natural History of Milan (MSNM) in close coordination with the Mediterranean Marine Mammals Tissue Bank (BTMM http://www.marinemammals.eu (accessed on 31 December 2020)) and the Cetacean stranding Emergency Response Team (CERT) of the University of Padova, which was also established with a mandate by the Ministry of Environment.

3. The Study Area

The Mediterranean Sea covers an area of 2.5 million km^2. Twenty-one countries and three different continents—Africa, Asia, and Europe—are affected by its waters. The Mediterranean basin communicates with the Atlantic Ocean and the Indo-Pacific area, respectively, through the Strait of Gibraltar and the Suez Canal, which are corridors that guarantee the passage of different cetacean species, predominantly from the west and less between the Mediterranean and the Red Sea, the latter arising for sporadic cases such as a specimen of *Sousa chinensis* [20].

From an oceanographical point of view, the Mediterranean Sea is divided into two macro-sectors: the western Mediterranean basin, which includes the Algerian-Provençal area and the Tyrrhenian, whose depth is no more than 3000 m in the northern Tyrrhenian; and the eastern Mediterranean basin, consisting of the Ionian, Aegean, and Levantine sub-basins, with a depth exceeding 5100 m in the Ionian area. The eastern Mediterranean basin is connected to the Black Sea through the straits of the Bosphorus and the Dardanelles. The Italian peninsula extends in the middle of the Mediterranean Sea and borders the two macro-sectors, connected by the Channel of Sicily (Figure 1).

Sicily Island is the southernmost region of Italy, has a coastline that is 1652 km long, 20% of the Italian coastal length, and overlooks three Mediterranean sub-basins (the Tyrrhenian, the Ionian Sea, the Channel of Sicily; Figure 1), each with peculiar oceanographical and ecological features.

Considering these aspects and its central position in the Mediterranean Sea, the study of marine mammals strandings in Sicily represents a topic of particular importance in the perspective of an effective network for both scientific research and museum enhancement. The position of Sicily is strategic, as its morphology makes the island a sort of *sentinel territory* for most marine species [21], due to the three portions of coast facing three different hydrographic and biogeographical provinces [22], i.e., the southern Tyrrhenian, along the northern coast, the Ionian, along the eastern coast, and the Channel of Sicily, along the southern coast.

The Tyrrhenian sub-basin is the deepest area in the western Mediterranean Sea [23]. It is characterized by complex bathymetry and plays an important role in the Mediterranean circulation because of several water masses flowing through [23]. The Atlantic Water (AW) enters the southern Tyrrhenian sub-basin in the upper layer of the water column (100–200 m thick); below the AW the Western Intermediate Water (WIW) is generated during the winter, while the West Mediterranean Deep Water (WMDW) flows at a greater depth. The complex dynamics and the presence of vortex and gyre structures are suitable conditions for vertical turbulence [24], resulting in favorable trophic conditions for cetaceans crossing the submarine canyons [25].

The Ionian Sea, the deepest regional area of the whole Mediterranean Sea, plays an important role in the intermediate and deep thermohaline cell of the Eastern Mediterranean conveyor belt. The Atlantic Water enters the Ionian, propagates towards the Levantine basin and bifurcates northward. Dense and oxygenated waters of Adriatic origin spread into the Ionian bottom layer, whilst the intermediate layer is influenced by salty and warm waters coming from the east [26]. Consequently, the Ionian circulation redistributes the different water masses rich in nutrients to adjacent seas. The Ionian continental shelf is very narrow, the depth along the eastern Sicilian coast drops suddenly reaching −2000 m within a few miles from the coastline, in contrast with the coastal seafloor morphology of the Channel of Sicily. The Ionian is characterized by significant upwellings that guarantee the regular sightings of six species: *Grampus griseus, Physeter macrocephalus, Balaenoptera physalus, Stenella coeruleoalba, Delphinus delphis,* and *Tursiops truncatus*, especially in the Gulf of Catania and in the strait of Messina [25,26].

The Channel of Sicily is a topographically complex region of the central Mediterranean comprising two sills: the depth of the eastern sill is about 540 m and that of the western is 530 m [27,28]. The maximum depth reaches 1700 m. The thermohaline circulation is mainly driven by an eastward flow of low-salinity Atlantic water (AW), bifurcating in the Atlantic Tunisian Current (ATC) and the Atlantic Ionian Stream (AIS), the last bordering the Sicilian coast. The AIS forces upwelling on the two shallow areas, the Adventure Bank and the Malta Plateau, influencing the concentration and distribution patterns of fish biomass, which is particularly favorable for the resident population of the common bottlenose dolphin, *Tursiops truncatus* [27–33].

The characteristics of the Mediterranean, in particular temperature and productivity (i.e., the presence of fish, macro-plankton, and cephalopods), affect the distribution of cetacean species [30,31]. Of the 78 known species, 22 have been recorded in the Mediterranean basin. The last species recorded in a recent stranding event is the first record of the Bryde's whale, *Balaenoptera edeni* (following nomenclature according to Kato and Perrin 2018 [34]), along the Egyptian coast [35].

The species of cetaceans observed in the Mediterranean Sea can be included into three categories [9,15,19,36]. Regular species, with resident populations, comprise 10 species including one belonging to the suborder Mysticeti (the fin whale, *Balaenoptera physalus*) and nine to suborder Odontoceti (the sperm whale, *Physeter macrocephalus*; the Cuvier's beaked whale, *Ziphius cavirostris*; the long-finned pilot whale, *Globicephala melas*; the Risso's dolphin, *Grampus griseus*; the common bottlenose dolphin, *Tursiops truncatus*; the striped dolphin, *Stenella coeruleoalba*; the short-beaked common dolphin, *Delphinus delphis*; and the Indo-Pacific rough-toothed dolphin, *Steno bredanensis*, which has only been observed in the Levantine Basin). Regarding *Steno bredanensis*, it is noteworthy to highlight that it has only recently been included as a regular species and maybe a relict population in the eastern basin [11]. The killer whale *Orcinus orca* can be also considered a regular species, resident in the Strait of Gibraltar whose population presence is widely verified by sightings [37]. Visitor species are named because of their Atlantic origin and have occasional appearances especially in the western Mediterranean basin (the false killer whale *Pseudorca crassidens*, the common minke whale *Balaenoptera acutorostrata*, and the humpback whale *Megaptera novaeangliae*). Vagrant species are those observed sporadically in different areas of the Mediterranean basin (the dwarf sperm whale *Kogia sima*, the northern bottlenose whale *Hyperoodon ampullatus*, the Blainville's beaked whale *Mesoplodon densirostris*, the Gervais' beaked whale *Mesoplodon europaeus*, the sei whale *Balaenoptera borealis*, the North Atlantic right whale *Eubalaena glacialis*, and the gray whale *Eschrichtius robustus*). Further, the Indo-Pacific humpback dolphin *Sousa chinensis* was included in a fourth category named alien species, the ones that moved towards the Mediterranean a few times following the opening of the Suez Canal (1869) (Morzer Bruyns, *pers. comm.* in Marchessaux, 1980) [19].

The conservation status of cetaceans in the Mediterranean is considered worrying by the IUCN (International Union for the Conservation of Nature), which draws up the "Red List of Threatened Species", the largest database of information on the conservation status of animal and plant species in all over the globe. Of the nine species of the Mediterranean cetaceans, *Ziphius cavirostris*, *Globicephala melas*, *Grampus griseus*, and *Steno bredaniensis* fall into the "Data Deficient" category; *Stenella coeruleoalba*, *Balaenoptera physalus*, and *Tursiops truncatus* fall into the "Vulnerable" category; *Delphinus delphis* and *Physeter macrocephalus* are instead considered to be "Endangered" [38].

4. The Census

In this work, the number of stranding events occurring throughout the national maritime zone, as recorded by the National Strandings Data Bank, was examined to relate it to those occurring along the coast of Sicily. The period examined was between 1990 and 2019. The Data Bank (http://mammiferimarini.unipv.it (accessed on 31 December 2020)) was consulted on 31 December 2019; any current discrepancy might derive from updates inserted after the consultation. We first proceeded by observing all records in each Italian administrative region, including stranded carcasses and entangled specimens (see Supplement Figure S1).

Figure 2 shows the data relating to the number of stranding events and the number of specimens (as a single event can include more than one specimen) and the number of species stranded for each Italian region. It should be taken into account that the effort for monitoring and reporting of strandings in the past was not homogenous across years, although this did not limit an accurate evaluation of strandings. A total of 4889 stranding events and 4970 specimens were counted.

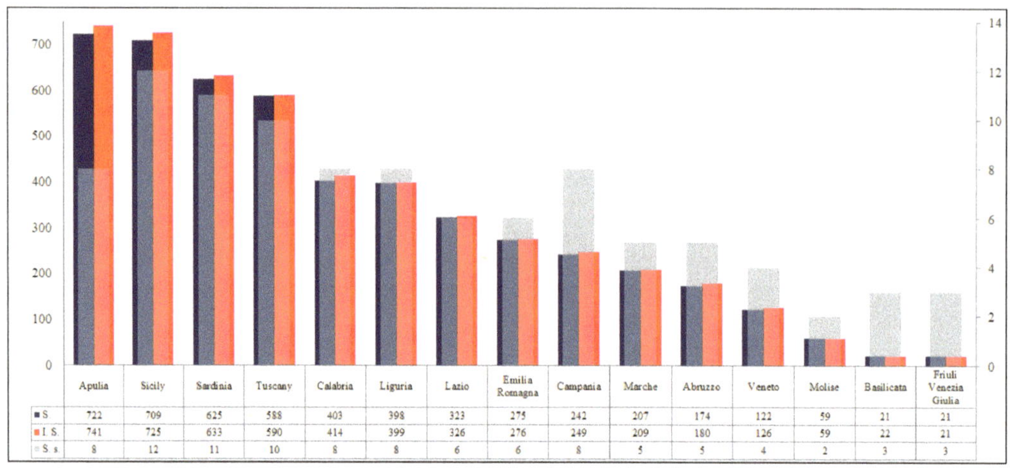

Figure 2. Number of stranding events (S.), the number of individuals stranded (I.S.), and the number of species stranded (S. s.) for each Italian region (see text for details and Supplement Figure S1 for map of the regions).

The extent of strandings appears proportionate to the length of the coastline per region (see Figure S1 in Supplementary material). Table 1 shows the number of specimens stranded, corresponding to the individuals stranded (I.S.) values of Figure 2, per species (1990–2019) in the different Italian regions.

Table 1. Number of specimens stranded, including entangled specimens (i.e., the I.S. in Figure 2) for each species (1990–2019) in the different Italian regions (Apu = Apulia; Sic = Sicily; Sar = Sardinia; Tus =Tuscany; Cal = Calabria; Lig = Liguria; Laz = Lazio; Emi = Emilia Romagna; Cam = Campania; Mar = Marche; Abr = Abruzzo; Ven = Veneto; Mol = Molise; Bas = Basilicata; Fri = Friuli Venezia Giulia) (see Supplemental Figure S1 for a map of the regions).

Species	Apu	Sic	Sar	Tus	Cal	Lig	Laz	Emi	Cam	Mar	Abr	Ven	Mol	Bas	Fri
Balaenoptera acutorostrata Lacépède, 1804		1	1	2		1				1					
Balaenoptera physalus (Linnaeus, 1758)	3	5	15	18	7	15	4	1	6	1					
Delphinus delphis Linnaeus, 1758	2	20	17	1				1	2	3	1	1			
Globicephala melas (Traill, 1809)	2	6	15	5	5	14									
Grampus griseus (G. Cuvier, 1812)	19	20	23	12	17	15	7	3	7	2	3	2			3
Kogia sima (Owen, 1866)		1							1						
Megaptera novaeangliae (Borowski, 1781)		1													
Mesoplodon europaeus (Gervasis, 1855)				1											
Physeter macrocephalus Linnaeus, 1758	10	45	29	9	18	4	16	1	14			7			
Pseudorca crassidens (Owen, 1846)			1		1										
Stenella coeruleoalba (Meyen, 1833)	318	335	282	292	267	240	185	12	129	4	29	9	5	11	2
Steno bredanensis (G. Cuvier, 1828)		6	1												
Tursiops truncatus (Montagu, 1821)	189	78	179	180	20	42	46	225	33	143	129	94	33	4	11
Ziphius cavirostris G. Cuvier, 1823	9	21	4	7	13	8	3							3	
Unidentified	189	186	66	63	66	60	65	33	56	56	11	20	21	4	5

The Italian regions where most of the strandings occurred are Apulia (Apu), Sicily (Sic), Sardinia (Sar), and Tuscany (Tus), counting respectively 722, 709, 625, and 588 stranding events, with, respectively, 741, 725, 633, and 590 specimens. The regions Molise (Mol), Basilicata (Bas), and Friuli Venezia Giulia (Fri) with the lowest coastline length showed a lower number of stranded cetaceans.

The common bottlenose dolphin, *Tursiops truncatus*, and the striped dolphin, *Stenella coeruleoalba*, are the most frequent species stranded in all regions. Sicily, compared to Apu, Sar, and Tus, has the lowest rate of strandings of *Tursiops truncatus*, and lists a greater variety of species along the coast, due to the *Balaenoptera acutorostrata*, *Kogia sima*, *Megaptera novaeangliae*, and *Steno bredanensis* occasional stranding events.

Regarding the two regions with the highest rate of strandings (Apu and Sic), important data concerns the Unidentified category, representing 25.5% and 25.6% of the total number of specimens recorded, respectively (Table 1).

This biological material, especially in Sicily, seems to be overlooked. An appropriate management program could retrieve it and contribute to increasing knowledge about cetacean distribution, increasing the valuable material for research and museum collections.

Sicily, occupying a central position in the Mediterranean, represents an area of transit of the more or less regular, sporadic, and vagrant species (Figure 3) whose presence and density is only documented through strandings, such as the dwarf sperm whale, *Kogia sima*, beached in Eraclea Minoa locality (AG) in June 2002 [39,40], the humpback whale, *Megaptera novaeangliae*, found entangled close to Siracusa and then released in 2004 [41], or the six specimens of the rough-toothed dolphin, *Steno bredanensis*, stranded along the Ionian coast of Sicily (RG) in April 2002, three of which died [40,42].

Figure 3. Some stranding events on the Sicilian coast: (**a**) *Stenella coeruleoalba*, striped dolphin, stranded in October 2018 (Triscina, Trapani); (**b**) *Physeter macrocephalus*, sperm whale, stranded in May 2019 (Capo Calavà, Gioiosa Marea, Messina); (**c**) *Grampus griseus*, Risso's dolphin, stranded in February 2020 (Milazzo, Messina); (**d**) *Tursiops truncates*, common bottlenose dolphin, stranded in January 2021 (Palermo) (photos ©Andrea Calascibetta); (**e**) Necroscopy procedure performed by a IZS and IAS-CNR Researchers Team for a specimen of *Balaenoptera physalus*, fin whale, stranded in September 2014 (Triscina di Selinunte, Trapani) (Photo © Giuseppa Buscaino, Bioacoustics Lab of CNR-IAS).

In addition to the reports of the aforementioned species, it is relevant to mention an old stranding event of false killer whales, *Pseudorca crassidens*, which occurred on the western Sicilian coast in 1877, before the analyzed period [43].

In the period 1990–2019, 725 specimens were counted along the Sicilian coastline (Figure 4). The greatest number of strandings (n = 106) occurred in the year 1991, which was due to a *Morbillivirus* infection [44]. The individuals concerned were two long-finned pilot whales, three sperm whales, 44 striped dolphins, five bottlenose dolphins, and 52 specimens that could not be identified as a result of decomposition. A decrease in numbers between 2005 and 2012 was probably due to lower efficiency in the monitoring of the region.

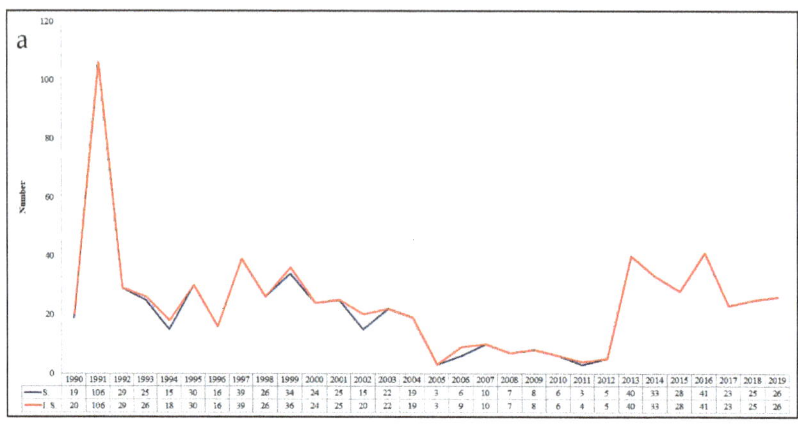

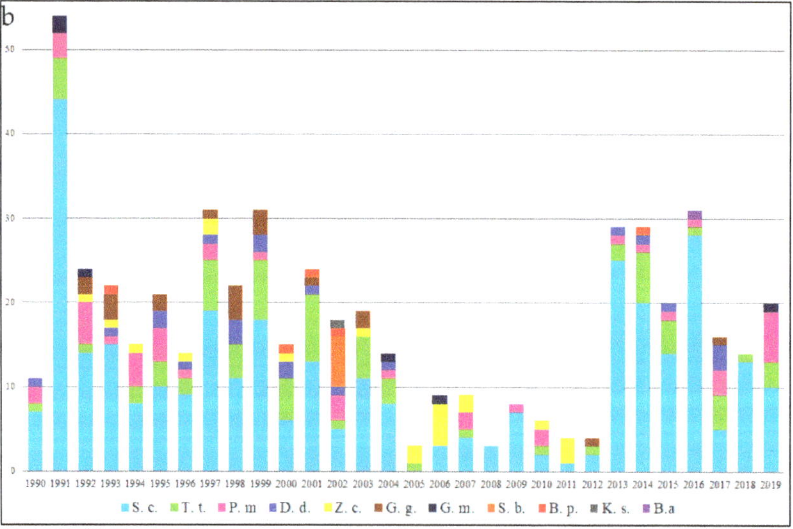

Figure 4. (**a**) Stranding frequency on Sicilian coast (1990–2019); S., Number of stranding events; I. S., number of stranded and entangled individuals; unidentified specimens included. (**b**) Number of stranded and entangled individuals for each species along the Sicilian coast per year (1990–2019). From the most to the less abundant *S. c.* = *Stenella coeruleoalba*; *T. t.* = *Tursiops truncatus*; *P. m.* = *Physeter macrocephalus*; *D. d.* = *Delphinus delphis*; *Z. c.* = *Ziphius cavirostris*; *G. g.* = *Grampus griseus*; *G. m.* = *Globicephala melas*; *S. b.* = *Steno bredanensis*; *B. p.* = *Balaenoptera physalus*; *K. s.* = *Kogia sima*; *M. n.* = *Megaptera novaeangliae*; *B. a.* = *Balaenoptera acutorostrata*; unidentified specimens not included.

On the whole, the annual average number of stranding events is 21, calculated by eliminating the maximum value (n = 106) and the minimum value (n = 3).

Stranding events showed a spatial pattern along the coast, as already detected in different seas [5–8], due to factors such as the seafloor morphology and the oceanographic characteristics. Table 2 shows the number of stranded individuals per sub-basin: Tyrrhenian, Ionian and Channel of Sicily. The most numerous strandings occurred in the Tyrrhenian sub-basin (n = 312), followed by the Channel of Sicily (n = 220), whereas the lowest number was scored in the Ionian sub-basin (n = 193).

Table 2. Number of stranded and entangled specimens for each species in the different Sicilian coastal areas (Tyrrhenian, Ionian, and Channel of Sicily).

Species	English Common Name	Italian Common Name	Tyrrhenian	Ionian	Channel of Sicily	Total
Balaenoptera acutorostrata	Common minke whale	Balenottera minore	1			1
Balaenoptera physalus	Fin whale	Balenottera comune	2		3	5
Delphinus delphis	Short-beaked common dolphin	Delfino comune	6	4	10	20
Globicephala melas	Long-finned pilot whale	Globicefalo	2		4	6
Grampus griseus	Risso's dolphin	Grampo	7	4	9	20
Kogia sima	Dwarf sperm whale	Cogia di Owen			1	1
Megaptera novaeangliae *	Humpback whale	Megattera		1		1
Physeter macrocephalus	Sperm whale	Capodoglio	31	4	10	45
Stenella coeruleoalba	Striped dolphin	Stenella striata	156	109	70	335
Steno bredanensis **	Rough-toothed dolphin	Steno			6	6
Tursiops truncatus	Common bottlenose dolphin	Tursiope	16	6	56	78
Ziphius cavirostris	Cuvier's beaked whale	Zifio	9	5	7	21
Unidentified			82	60	44	186
Total			312	193	220	725

* entangled and then released; ** all stranded and three released.

Figure 5 shows the different percentages of the species stranded in the different Sicilian coastal sectors. In particular, it should be noticed that in the Tyrrhenian sub-basin the most frequent species are the striped dolphin *Stenella coeruleoalba* (50%), the sperm whale *Physeter macrocephalus* (10%), and the common bottlenose dolphin *Tursiops truncatus* (5%); in the Ionian coast, the most frequent species are the striped dolphin *Stenella coeruleoalba* (56%), the common bottlenose dolphin *Tursiops truncatus*, and the Cuvier's beaked whale *Ziphius cavirostris* (3%); in the Channel of Sicily, the most frequent species are the striped dolphin *Stenella coeruleoalba* (32%), the common bottlenose dolphin *Tursiops truncatus* (25%), the fin whale *Balaenoptera physalus*, and the sperm whale *Physeter macrocephalus* (5%). Regarding the unidentified carcasses, 26% were in the first sector, 30% in the second, and finally 20% in the third, respectively.

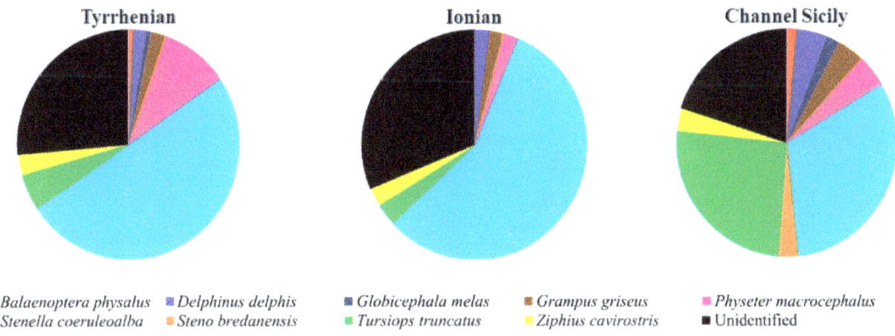

Figure 5. Percentage of carcasses of all stranded species during the period 1990–2019 (see Table 2) per each Sicilian coastal sub-area (Tyrrhenian, Ionian, and Channel of Sicily), unidentified specimens included, unique specimens excluded.

5. Sicily as a Crossroads for Cetaceans' Passage

The results highlight that the Tyrrhenian sub-basin, with 312 specimens, shows the highest number of stranded specimens, followed by the Channel of Sicily (n = 220) and the Ionian (n = 193); whereas the coast that receives the highest number of species is the Channel of Sicily (10 species), followed by the Tyrrhenian sub-basin (9 species) and the Ionian sub-basin (7 species).

On the Sicilian coast, 12 stranded species were counted of the 22 species of cetaceans that have been reported in the Mediterranean Sea. This data remarks the value of Sicily in common and rare cetacean species recruitment in the Mediterranean Sea and confirms the importance of the central position in the basin as a crossroads for cetacean passage.

Different factors, such as population density, the distance between the site of death and the coastline, the buoyancy of the carcasses, winds, and currents, can determine the number of stranding records. Considering the greatly extended coastline of the three Sicilian sub-areas, cetacean stranding records can reflect the relative abundance of living populations inhabiting the neighbouring areas [5,45] and could be a good source of information when survey efforts at sea are scarce or absent [8,46].

The cetacean stranding records can be moreover affected by variation in reporting rates by "citizen science" activities [8,30,47] and by the presence of a stranding network to collect and validate datasets [45,48].

The differences in abundance and the species diversity resulting in this study are correlated to different environmental and anthropogenic features. The three Sicilian coastal sub-areas show that different bathymetry, ecological characteristics, naval traffic density and fisheries influence the distribution, the behavior and the life safety of cetaceans. Our analysis confirms that *Stenella coeruleoalba* and *Tursiops truncatus* are the most commonly found species around Sicily [19,49], and similarly notes some differences in the distribution of them among the three sub-basins due to the environmental and anthropogenic features described above.

Regarding the *Stenella coeruleoalba* population, the number of strandings is the lowest in the Channel of Sicily sub-area; in this case, due to the large and shallow continental shelf, the populations live far from the coast and dead specimens can float offshore towards different areas of the Mediterranean. Regarding *Tursiops truncatus*, data indicate the Channel of Sicily as the sub-basin in which most specimens are stranded. This data is coherent with the nearshore habitat use of the common bottlenose dolphin [8,50]. The high number of stranded bottlenose dolphins adequately reflects previous studies of its population and interactions with industrial fishing activities [33,49].

Additionally, the distribution of the strandings is different regarding the sperm whale, *Physeter macrocephalus*, which strands more frequently on the Tyrrhenian coast. Previous literature reports the presence of this species in the Channel of Sicily, in the Ionian Sea [51] and in the Tyrrhenian Sea, commonly concentrated in canyon areas [52]. The present results show its strandings somehow infrequent along the eastern Sicilian coast, where the Ionian seafloor morphology characterized by a wide abyssal plain can limit risks of death.

It should be noticed that though the short-beaked common dolphin, *Delphinus delphis*, which was formerly very common and successively undergone a dramatic decline in abundance during the last few decades [19], shows very rare stranding events within the 30-year period, both in Italy and Sicily, if compared with other species like *Grampus griseus* or *Physeter macrocephalus*.

Finally, an interesting fact concerns the six records of *Steno bredanensis* stranded in the Sicilian Channel because the only sightings of this species have been made in the eastern Mediterranean [40,42] and in the Tyrrhenian sub-basin, in the Lazio (Laz) area [25].

6. The Cetological Collections in Sicily

From earlier literature (see Supplement Document F1), it emerges that Sicily is the seventh region in Italy with regard to the number of cetological collections (88 records), but it exhibits a small number of complete skeletons to the public.

The number of collected cetaceans in Sicily could have been higher if the local institutions would have been better organized in recovering carcasses.

Several Italian museums not located in Sicily store specimens (24) collected from the Sicilian coasts (in Milan [53]; Florence [54]; Pisa [16]; Genoa [55]; Padua [56]; Livorno [57]). The collections of cetaceans in Sicily are mainly osteological [12,14] and preserve a few specimens distributed in several exhibitions. The museum with the largest collection is the Civic Museum of Natural History of Comiso which stores 39 pieces obtained from specimens collected from 1991 to 2003 in Sicily. Among the specimens of considerable relevance, a complete and disjointed skeleton of the dwarf sperm whale, *Kogia sima*, the sole individual ever stranded in Sicily, and two skeletons of the rough-toothed dolphin, *Steno bredanensis* [12].

The Museum of Zoology "Pietro Doderlein" of the University of Palermo preserves the second most important cetological collection, albeit consisting of only 18 pieces (Figure 6a): two short-beaked common dolphins, *Delphinus delphis*, taxidermied; two common bottlenose dolphin, *Tursiops truncatus*, skulls; two Risso's dolphin, *Grampus griseus*, skulls; one Cuvier's beaked whale, *Ziphius cavirostris*, skull; one short-beaked common dolphin, *Delphinus delphis*, skull, and a partial skeleton. The specimens collected have a historical value, including a partial skeleton of a sperm whale, *Physeter macrocephalus*, that beached alive near the Stagnone of Marsala (TP) in December 1872, in a massive stranding event [58]. Other unexposed specimens are two fetuses of short-beaked common dolphins, *Delphinus delphis*, and two fin whale, *Balaenoptera physalus*, vertebrae.

Other Sicilian museums that house cetological collections are the Museo di Zoologia e Casa delle Farfalle of the University of Catania (4 specimens), the Museo della Fauna of the University of Messina (2 specimens), the Zoological Museum "F. Cambria" of the University of Messina (3 specimens), and the Acquario Civico in Messina (7 specimens) [12].

A cetological reconstruction laboratory established by the Institute of Anthropic Impact and Sustainability in Marine Environment of the National Research Council (CNR-IAS) at Capo Granitola (Sicily) collaborates with national and international experts with the aim to fill the skills gap on museological preparations and enhance and create a collection of skeletal systems of Mediterranean cetaceans, which is accessible to research and available for scientific dissemination. In the laboratory, several complete skeletons (*Balaenoptera physalus*, *Ziphius cavirostris*, *Tursiops truncatus*, two *Stenella coeruleoalba* specimens), *Delphinus delphis* (2 specimens) and incomplete skeletal and skulls are stored. Actually, one skeleton of *Physeter macrocephalus* (Figure 6b), one of *Stenella coeruleoalba* and one of *Grampus griseus* are included as part of a permanent exhibition of the "Observatory of terrestrial and marine Biodiversity of the Sicilian Region" (ORBS).

The extent of the cetological collections exhibited in Sicilian museums does not depend on a lack of available resources as the island is an area where numerous strandings occur. On the contrary, this is due to the loss of 93.5% of the specimens stranded in Sicily, which are not recovered by museums but instead discarded.

There are several reasons for this. First of all, there is an absence of some skills such as taxidermy due to the profession no longer being of interest for young people, the lack of financial resources dedicated to this field, and the limited interest of several public institutions. However, Sicily in recent years has received great consideration by the scientific authorities regarding the phenomenon of strandings and the recovery of the skeletons for museum purposes.

In September 2014 a 20 m-long fin whale, *Balaenoptera physalus*, a decomposed specimen, stranded close to Triscina di Selinunte, Castelvetrano (Trapani), whose analyses were managed by CNR-IAS and IZS (Istituto Zooprofilattico della Sicilia) researchers (Figure 3). The skeleton was extracted and is now preserved for research and scientific dissemination activities at the CNR-IAS of Capo Granitola (Gaspare Buffa *pers. comm.*). This stranding event is noteworthy as the fin whale is the largest species ever recorded along the island's coast. It should be noticed that a mistake was reported in the BDS and a 12 m-long *Balaenoptera physalus* specimen was archived.

Figure 6. Examples of Sicilian cetological collections. (**a**) A view of the exhibition at the Museum of Zoology "P. Doderlein" of the University of Palermo. From above to below: *Delphinus delphis* skull, partial skeleton and anatomy apparatus; two *Tursiops truncatus* skulls, and a *Stenella coeruleoalba* skull; two *Grampus griseus* skulls. (**b**) The *Physeter macrocephalus* skeleton at the Institute of Anthropic Impact and Sustainability in marine Environment (CNR-IAS) section of Capo Granitola, Campobello di Mazara (Tp) (photos ©Andrea Calascibetta and © Gaspare Buffa).

In October 2016, an 8.4 m-long sperm whale, *Physeter macrocephalus*, stranded along the coast of Aspra (Palermo) was recovered by the task force of CERT (Cetacean Stranding Emergency Response Team-UNIPD) on behalf of the local authority to carry out the reconstruction of the skeleton (Sandro Mazzariol *pers.comm.*). In 2017, a stranded sperm whale was recovered by the Museo della Fauna (University of Messina) and exhibited at the Castle of Milazzo (Messina). Further, in February and May 2019, a long-finned pilot whale, *Globicephala melas*, stranded in Barcellona Pozzo di Gotto (Messina), and a sperm whale found in Cefalù (Palermo) were also recovered by the same museum (Filippo Spadola *pers.comm.*). Thus, an increasing interest in building a strandings network enables sharing skills and information and improving the samples collection.

7. Conclusions

In light of the aforementioned discussion, we can assert that Sicily is an island with a potential cetological resource that can integrate a museum heritage at a national and international scale. Under the sustainable development goals (SDGs) targeted by the United Nations [59] which include relevant topics on the protection and conservation of marine life, cetacean strandings can become a tool for the implementation of scientific dissemination programs if carcasses are adequately recovered and preserved. This paper aims at promoting the establishment of a network for the management of strandings, the training of specialized personnel, and the collaboration among researchers to better improve knowledge about these marine mammals and the ocean environment.

Supplementary Materials: The following are available online at https://www.mdpi.com/1424-281 8/13/3/104/s1, Figure S1: Italian strandings during 1990–2019. Map of Italian regions and a table with the number of strandings (S) and the number of individuals stranded (I. S.) for each Italian region. Coastline length has been obtained from the Istat portal (https://www.istat.it/it/archivio/137341 (accessed on 31 December 2020)); Document F1: List of references reporting information about the inventory of the cetological collections in the Italian museums per region; Document F2: Annual reports published by the Natural History Museum of Milan (Atti Soc. Ital. Sci. nat. Museo civ. Stor. Nat. Milano from I of 1986 to XXI of 2012).

Author Contributions: Conceptualization, A.C, S.L.B., G.B., and G.P.; investigation, A.C.; resources, S.L.B.; original draft writing A.C., S.L.B., and G.P.; translation, S.L.B. and G.B.; final review and editing, S.L.B., G.B., G.P., and A.C. All authors have read and agreed to the published version of the manuscript.

Funding: This research was funded by the University of Palermo—FFR grant.

Acknowledgments: The authors are grateful to Filippo Spadola Director of the Museo della Fauna (Universiy of Messina), to Sandro Mazzariol (University of Padova) for having corroborated information about the new specimens recovered by public institutions, and to the three anonymous referees which suggested useful revisions. The authors are grateful also to Roberto Puleio, Domenico Vicari and Salvatore Seminara of the IZS (Istituto Zooprofilattico della Sicilia) for their efforts in managing and recovering stranded mammals along Sicilian coast. The role of IZS is preeminent in such purpose.

Conflicts of Interest: The authors declare no conflict of interest.

References

1. Rainbow, P.S. Marine biological collections in the 2tablet century. *Zool. Scr.* **2009**, *38*, 33–40. [CrossRef]
2. Gravina, M.F.; Bonifazi, A.; Del Pasqua, M.; Giampaoletti, J.; Lezzi, M.; Ventura, D.; Giangrande, A. Perception of changes in marine benthic habitats: The relevance of taxonomic and ecological memory. *Diversity* **2020**, *12*, 480. [CrossRef]
3. Mannino, A.M.; Balistreri, P.; Iaciofano, D.; Galil, B.S.; Lo Brutto, S. An additional record of *Kyphosus vaigiensis* (Quoy & Gaimard, 1825) (Osteichthyes, Kyphosidae) from Sicily clarifies the confused situation of the Mediterranean kyphosids. *Zootaxa* **2015**, *3963*, 45–54.
4. Smith, K.J.; Sparks, J.P.; Timmons, Z.L.; Peterson, M.J. Cetacean skeletons demonstrate ecologically relevant variation in intraskeletal stable isotopic values. *Front. Mar. Sci.* **2020**, *7*, 388. [CrossRef]
5. Foord, C.S.; Rowe, K.M.; Robb, K. Cetacean biodiversity, spatial and temporal trends based on stranding records (1920–2016), Victoria, Australia. *PLoS ONE* **2019**, *14*, e0223712. [CrossRef]
6. Pikesley, S.K.; Witt, M.J.; Hardy, T.; Loveridge, J.; Loveridge, J.; Williams, R.; Godley, B.J. Cetacean sightings and strandings: Evidence for spatial and temporal trends? *J. Mar. Biol. Assoc. UK* **2012**, *92*, 1809. [CrossRef]
7. Peltier, H.; Dabin, W.; Daniel, P.; Van Canneyt, O.; Dorémus, G.; Huon, M.; Ridoux, V. The significance of stranding data as indicators of cetacean populations at sea: Modelling the drift of cetacean carcasses. *Ecol. Indic.* **2012**, *18*, 278–290. [CrossRef]
8. Santos, M.C.D.O.; Siciliano, S.; Vicente, A.F.D.C.; Alvarenga, F.S.; Zampirolli, É.; Souza, S.P.D.; Maranho, A. Cetacean records along São Paulo state coast, southeastern Brazil. *Braz. J. Oceanogr.* **2010**, *58*, 123–142. [CrossRef]
9. Notarbartolo di Sciara, G. Marine mammals in the Mediterranean Sea: An overview. *Adv. Mar. Biol.* **2016**, *75*, 95–106.
10. Maio, N.; De Stasio, R. La collezione cetologica del Museo Zoologico dell'Università degli Studi di Napoli Federico II. Catalogo aggiornato e ragionato. *Museol. Sci. Mem.* **2014**, *12*, 327–342.
11. Kerem, D.; Goffman, O.; Elasar, M.; Hadar, N.; Scheinin, A.; Lewis, T. The Rough-Toothed—Dolphin, *Steno bredanensis*, in the Eastern Mediterranean Sea: A Relict Population? *Adv. Mar. Biol.* **2016**, *75*, 233–258.
12. Insacco, G.; Buscaino, G.; Buffa, G.; Cavallaro, M.; Crisafi, E.; Grasso, R.; Lombardo, F.; Lo Paro, G.; Parrinello, N.; Sarà, M.; et al. Il patrimonio delle raccolte cetologiche museali della Sicilia. *Museol. Sci. Mem.* **2014**, *12*, 391–405.
13. Baccetti, N.; Cancelli, F.; Renieri, T. First record of *Kogia simus* (Cetacea, Physeteridae) from the Mediterranean Sea. *Mammalia* **1991**, *55*, 152–154.
14. Cagnolaro, L.; Podestà, M.; Affronte, M.; Agnelli, P.; Cancelli, F.; Capanna, E.; Carlini, R.; Cataldini, G.; Cozzi, B.; Insacco, G.; et al. Collections of extant cetaceans in Italian museums and other scientific institutions. A comparative review. *Nat. Hist. Sci.* **2012**, *153*, 145–202. [CrossRef]

15. Notarbartolo di Sciara, G.; Bearzi, G. Research on cetaceans in Italy. In *Marine Mammals of the Mediterranean Sea: Natural History, Biology, Anatomy, Pathology, Parasitology*; Valdina, M., Ed.; The Coffee House Art Adv: Milano, Italy, 2005; pp. 1–25.

16. Nicolosi, P.; Braschi, S.; Cagnolaro, L.; Zuffi, M.A. Il patrimonio di Cetacei attuali del Museo di Storia naturale dell'Università di Pisa (Certosa di Calci). Profilo storico e catalogo della collezione. *Museol. Sci. Mem.* **2014**, *12*, 215–238.

17. Pavan, G.; Podestà, M.; D'Amico, A.; Portunato, N.; Fossati, C.; Manghi, M.; Priano, M.; Quero, M.; Teloni, V. A GIS and associated database for the Italian Stranding Network. A cooperative project based on GIS technologies. In *European Research on Cetaceans, Proceedings of the 16th ECS Conference, Liege, Belgium, 7–11 April 2002*; Evans, P.G.H., Lockyer, C.H., Buckingham, L., Jauniaux, T., Eds.; European Cetacean Society: Kiel, Germany, 2002; Volume 16, pp. 101–104.

18. Pavan, G.; Bernuzzi, E.; Cozzi, B.; Podestà, M. La rete nazionale di monitoraggio degli spiaggiamenti di mammiferi marini. *Biol. Mar. Mediterr.* **2013**, *20*, 262–263.

19. Notarbartolo di Sciara, G.; Venturino, M.C.; Zanardelli, M.; Bearzi, G.; Borsani, F.J.; Cavalloni, B. Cetaceans in the central Mediterranean Sea: Distribution and sighting frequencies. *Ital. J. Zool.* **1993**, *60*, 131–138.

20. Kerem, D.; Goffman, O.; Spanier, E. Sighting of a single humpback dolphin (*Sousa* sp.) along the Mediterranean coast of Israel. *Mar. Mamm. Sci.* **2001**, *17*, 170–171. [CrossRef]

21. Servello, G.; Andaloro, F.; Azzurro, E.; Castriota, L.; Catra, M.; Chiarore, A.; Crocetta, F.; D'Alessandro, M.; Denitto, F.; Froglia, C.; et al. Marine alien species in Italy: A contribution to the implementation of Descriptor D2 of the Marine Strategy Framework Directive. *Mediterr. Mar. Sci.* **2019**, *20*, 1. [CrossRef]

22. Bianchi, C.N.; Morri, C. Marine biodiversity of the Mediterranean Sea: Situation, problems and prospects for future research. *Mar. Pollut. Bull.* **2000**, *40*, 367–376. [CrossRef]

23. Budillon, G.; Gasparini, G.P.; Schroeder, K. Persistence of an eddy signature in the central Tyrrhenian basin. *Deep Sea Res. II* **1996**, *56*, 713–724. [CrossRef]

24. Astraldi, M.; Gasparini, G.P. The seasonal characteristics of the circulation in the Tyrrhenian Sea. In *Seasonal and Interannual Variability of the Western Mediterranean Sea, Coastal and Estuarine Studies*; AGU: Washington, DC, USA, 1994; Volume 46, pp. 115–134.

25. Santoro, R.; Sperone, E.; Tringali, M.L.; Pellegrino, G.; Giglio, G.; Tripepi, S.; And Arcangeli, A. Summer Distribution, Relative Abundance and Encounter Rates of Cetaceans in the Mediterranean Waters off Southern Italy (Western Ionian Sea and Southern Tyrrhenian Sea). *Mediterr. Mar. Sci.* **2015**, *16*, 613–620. [CrossRef]

26. Gacić, M.; Borzelli, G.L.E.; Civitarese, G.; Cardin, V.; Yari, S. Can internal processes sustain reversals of the ocean upper circulation? The Ionian Sea example. *Geophys. Res. Lett* **2010**, *37*, 5. [CrossRef]

27. Bonanno, A.; Placenti, F.; Basilone, G.; Mifsud, R.; Genovese, S.; Patti, B.; Di Bitetto, M.; Aronica, S.; Barra, M.; Giacalone, G.; et al. Variability of water mass properties in the Strait of Sicily in summer period of 1998–2013. *Ocean Sci.* **2014**, *10*, 759–770. [CrossRef]

28. Gasparini, G.P.; Ortona, A.; Budillon, G.; Astraldi, M.; Sansone, E. The effect of the Eastern Mediterranean Transient on the hydrographic characteristics in the Strait of Sicily and in the Tyrrhenian Sea. *Deep Sea Res. I* **2005**, *52*, 915–935. [CrossRef]

29. Placenti, F.; Schroeder, K.; Bonanno, A.; Zgozi, S.; Sprovieri, M.; Borghini, M.; Rumolo, P.; Cerrati, G.; Bonomo, S.; Genovese, S.; et al. Water masses and nutrient distribution in the Gulf of Syrte and between Sicily and Libya. *J. Marine Sys.* **2013**, *121*, 36–46. [CrossRef]

30. Pace, D.S.; Giacomini, G.; Campana, I.; Paraboschi, M.; Pellegrino, G.; Silvestri, M.; Alessi, J.; Angeletti, D.; Cafaro, V.; Pavan, G.; et al. An integrated approach for cetacean knowledge and conservation in the central Mediterranean Sea using research and social media data sources. *Aquat. Conserv. Mar. Freshw. Ecosyst.* **2019**, *29*, 1302–1323. [CrossRef]

31. Gannier, A. Summer distribution and relative abundance of delphinids in the Mediterranean Sea. *Rev. Ecol.* **2005**, *60*, 223–238.

32. Bellante, A.; Sprovieri, M.; Buscaino, G.; Buffa, G.; Di Stefano, V.; Salvagio Manta, D.; Barra, M.; Filicciotto, F.; Bonanno, A.; Mazzola, S. Distribution of Cd and as in organs and tissues of four marine mammal species stranded along the Italian coasts. *J. Environ. Monit.* **2012**, *14*, 2382. [CrossRef]

33. Papale, E.; Ceraulo, M.; Buffa, G.; Filicciotto, F.; Grammauta, R.; Maccarrone, V.; Mazzola, S.; Buscaino, G. Association patterns and population dynamics of bottlenose dolphins in the Strait of Sicily (Central Mediterranean Sea) implication for management. *Popul. Ecol.* **2017**, *59*, 55–64. [CrossRef]

34. Kato, H.; Perrin, W.F. *Bryde's Whale: Balaenoptera Edeni*; Würsig, B., Thewissen, J.G.M., Kovacs, K.M., Eds.; Academic Press: Cambridge, MA, USA, 2018; pp. 143–145.

35. Abo-Taleb, H.A.; El-feky, M.M.M.; El-Tabakh, M.A.M.; Hendy, D.M.; & Maaty, M. First record of Bryde's whale (*Balaenoptera brydei*, Olsen, 1913) in the southeastern Mediterranean Sea, Alexandria, Egypt. *Egypt. J. Aquat. Biol. Fish.* **2020**, *24*, 667–695. [CrossRef]

36. Loy, A.; Aloise, G.; Ancillotto, L.; Angelici, F.M.; Bertolino, S.; Capizzi, D.; Fontaneto, D. Mammals of Italy: An annotated checklist. *Hystrix* **2019**, *30*, 87–106.

37. Pace, D.S.; Tizzi, R.; Mussi, B. Cetaceans value and conservation in the Mediterranean Sea. *J. Biodiv. Endanger. Species* **2015**. [CrossRef]

38. IUCN. *Marine Mammals and Sea Turtles of the Mediterranean and Black Seas*; IUCN: Gland, Switzerland; Malaga, Spain, 2012.

39. Bortolotto, A.; Papini, L.; Insacco, G.; Gili, C.; Tumino, G.; Mazzariol, S.; Pavan, G.; Cozzi, B. First record of a dwarf sperm whale, *Kogia sima* (Owen, 1866) stranded alive along the coasts of Italy. In Proceedings of the 31st Symposium of the European Association for Aquatic Mammals, Tenerife, Spain, 14–17 March 2003.

40. Insacco, G.; Spadola, F.; Scaravelli, D.; Zava, B. Report on cetacean strandings in Sicily from 1991 to 2013. *Nat. Rerum* **2016**, *4*, 1–13.

41. Onlus, C.S.C.; Di Milano, M.C.D.S.N. Centro Studi Cetacei. Cetacei spiaggiati lungo le coste italiane. XIX. Rendiconto 2004 (Mammalia). *Atti Della Società Italiana di Scienze Naturali e del Museo Civico di Storia Naturale di Milano* **2006**, *147*, 145–157.

42. Watkins, W.A.; Tyack, P.; Moore, K.E.; Notarbartolo di Sciara, G. *Steno bredanensis* in the Mediterranean Sea. *Mar. Mamm. Sci.* **1987**, *3*, 78–82. [CrossRef]

43. Riggio, G. Sul *Globicephalus melas* Traill. *Nat. Sicil.* **1882**, *2*, 7–10.

44. Di Guardo, G.; Di Francesco, C.E.; Eleni, C.; Cocumelli, C.; Scholl, F.; Casalone, C.; Peletto, S.; Mignone, W.; Tittarelli, C.; Di Nocera, F.; et al. *Morbillivirus* infection in cetaceans stranded along the Italian coastline: Pathological, immunohistochemical and biomolecular findings. *Res. Vet. Sci.* **2013**, *94*, 132–137. [CrossRef] [PubMed]

45. Pyenson, N.D. The high fidelity of the cetacean stranding record: Insights into measuring diversity by integrating taphonomy and macroecology. *Proc. R. Soc. Lond. B Biol. Sci.* **2011**, *278*, 3608–3616. [CrossRef] [PubMed]

46. D'Astore, P.; Bearzi, G.; Bonizzoni, S. Cetacean strandings in the province of Brindisi (Italy, southern Adriatic Sea). *Ann. Ser. Hist. Nat.* **2008**, *18*, 29–38.

47. Authier, M.; Peltier, H.; Dorémus, G.; Dabin, W.; Van Canneyt, O.; Ridoux, V. How much are stranding records affected by variation in reporting rates? A case study of small delphinids in the Bay of Biscay. *Biodivers. Conserv.* **2014**, *23*, 2591–2612. [CrossRef]

48. Farrag, M.M.S.; Ahmed, H.O.; TouTou, M.M.M.; Eissawi, M.M. Marine Mammals on the Egyptian Mediterranean Coast "Records and Vulnerability". *Int. J. Ecotoxicol. Ecobiol.* **2019**, *4*, 8–16. [CrossRef]

49. Crosti, R.; Arcangeli, A.; Romeo, T.; Andaloro, F. Assessing the relationship between cetacean strandings (*Tursiops truncatus* and *Stenella coeruleoalba*) and fishery pressure indicators in Sicily (Mediterranean Sea) within the framework of the EU Habitats Directive. *Eur. J. Wildl. Res.* **2017**, *63*, 55. [CrossRef]

50. Akkaya, A.; Lyne, P.; Schulz, X.; Awbery, T.; Capitain, S.; Rosell, B.F.; Yıldırım, B.; İlkılınç, C.; Vigliano Relva, J.; Clark, H.; et al. Preliminary results of cetacean sightings in the eastern Mediterranean Sea of Turkey. *J. Black Sea/Medit. Environ.* **2020**, *26*, 26–47.

51. Lewis, T.; Gillespie, D.; Lacey, C.; Matthews, J.; Danbolt, M.; Leaper, R.; McLanaghan, R.; Moscrop, A. Sperm whale abundance estimates from acoustic surveys of the Ionian Sea and Straits of Sicily in 2003. *J. Mar. Biol. Assoc. U.K.* **2007**, *87*, 353–357. [CrossRef]

52. Mussi, B.; Miragliuolo, A.; Zucchini, A.; Pace, D.S. Occurrence and spatio-temporal distribution of sperm whale (*Physeter macrocephalus*) in the submarine canyon of Cuma (Tyrrhenian Sea, Italy). *Aquat. Conserv. Mar. Freshw. Ecosyst.* **2014**, *24* (Suppl. 1), 59–70. [CrossRef]

53. Podestà, M.; Bardelli, G.; Cagnolaro, L. Catalogo dei cetacei attuali del Museo di Storia Naturale di Milano. *Museol. Sci. Mem.* **2014**, *12*, 24–51.

54. Agnelli, P.; Ducci, L.; Funaioli, U.; Cagnolaro, L. La collezione dei Cetacei attuali del Museo di Storia Naturale dell'Università di Firenze: Indagine storica e revisione sistematica. *Museol. Sci. Mem.* **2014**, *12*, 194–212.

55. Poggi, R. I Cetacei del Museo Civico di Storia Naturale "Giacomo Doria" di Genova. *Museol. Sci. Mem.* **2014**, *12*, 117–152.

56. Nicolosi, P. I cetacei del Museo di Zoologia dell'Università di Padova. *Museol. Sci. Mem.* **2014**, *12*, 88–91.

57. Roselli, A.; Borzatti von Löwenstern, A.; Bisconti, M. La collezione osteologica di cetacei del Museo 300 di Storia Naturale del Mediterraneo della Provincia di Livorno. *Museol. Sci. Mem.* **2014**, *12*, 239–248.

58. Riggio, G. Arenamento di sette capidogli (*Physeter (Catodon) macrocephalus*, Lin.) nel mare di Marsala. *Nat. Sicil.* **1893**, *12*, 103–108.

59. Messerli, P.; Murningtyas, E.; Eloundou-Enyegue, P.; Foli, E.G.; Furman, E.; Glassman, A.; Hernández Licona, G.; Kim, E.M.; Lutz, W.; Moatti, J.P. Independent Group of Scientists appointed by the Secretary-General. In *Global Sustainable Development Report 2019: The Future Is Now—Science for Achieving Sustainable Development*; United Nations publication issued by the Department of Economic and Social Affairs: New York, NY, USA, 2019.

Opinion

Perception of Changes in Marine Benthic Habitats: The Relevance of Taxonomic and Ecological Memory

Maria Flavia Gravina [1,2,*], Andrea Bonifazi [1], Michela Del Pasqua [3], Jacopo Giampaoletti [1], Marco Lezzi [4], Daniele Ventura [5] and Adriana Giangrande [2,6]

1   Department of Biology, University of Rome "Tor Vergata", 00133 Rome, Italy; bonifazi.andrea@virgilio.it (A.B.); jacopogiampaoletti@gmail.com (J.G.)
2   CoNISMa, Consorzio Interuniversitario per le Scienze del Mare, 00196 Rome, Italy; adriana.giangrande@unisalento.it
3   ARPAE, Regional Agency for Environmental Prevention and Energy of Emilia Romagna, 48121 Ravenna, Italy; michela.delpasqua@unisalento.it
4   ARPAE, Regional Agency for Prevention, Environment and Energy, Emilia-Romagna, 47042 Cesenatico, Forlì-Cesena, Italy; marcolezzi86@gmail.com
5   Department of Environmental Biology, University of Rome "Sapienza", 00185 Rome, Italy; daniele.ventura@uniroma1.it
6   Department of Biological and Environmental Sciences and Technologies, University of Salento, 73100 Lecce, Italy
*   Correspondence: maria.flavia.gravina@uniroma2.it

Received: 27 October 2020; Accepted: 14 December 2020; Published: 16 December 2020

Abstract: Having a reliable ecological reference baseline is pivotal to understanding the current status of benthic assemblages. Ecological awareness of our perception of environmental changes could be better described based on historical data. Otherwise, we meet with the shifting baseline syndrome (SBS). Facing SBS harmful consequences on environmental and cultural heritage, as well as on conservation strategies, requires combining historical data with contemporary biomonitoring. In the present "era of biodiversity", we advocate for (1) the crucial role of taxonomy as a study of life diversity and (2) the robust, informative value of museum collections as memories of past ecosystem conditions. This scenario requires taxonomist skills to understand community composition and diversity, as well as to determine ecosystem change trends and rates. In this paper, we focus on six Mediterranean benthic habitats to track biological and structural changes that have occurred in the last few decades. We highlight the perception of biological changes when historical records make possible effective comparisons between past reference situations and current data. We conclude that the better we know the past, the more we understand present (and will understand future) ecosystem functioning. Achieving this goal is intrinsically linked to investing in training new taxonomists who are able to assure intergeneration connectivity to transmit cultural and environmental heritage, a key aspect to understanding and managing our changing ecosystems.

Keywords: environmental changes; biodiversity; historical data; benthic communities; biomonitoring; taxonomy; museum collections; environmental heritage

---

1. Changes in Marine Benthic Communities

Πάντα ῥεῖ, (panta rei), the well-known Heraclitus aphorism, admirably highlights change as the manifold aspect of Nature. Everything flows, as "one cannot descend twice into the same river and one cannot touch twice a substance in the same state". Environmental space and time changes continuously occur, trigging variations in organism assemblages, whose description is the major goal of ecology. Both irreversible (evolutionary approach) and reversible (ecological approach) changes characterize life history. Even in the ecological perspective, variations may be investigated at both synchronic (biological and physical–chemical differences in space) and diachronic (modifications in a temporal scale) levels.

Ecosystem dynamics have been the favorite topic among marine ecologists, including succession, persistence, and evolution, and the Mediterranean benthic communities have not been an exception (e.g., [1–3]). Such an interest is according to its relevance in resource partitioning processes. Indeed, successful strategies are linked to species alternation in biomass contribution through time in response to the adaptation to environmental changes and resource availability, as described by the "flush and crash" model [4]. Understanding changes also means distinguishing predictable and unpredictable modifications, as well as periodical/progressive fluctuations, from sudden/short-term variations. The manifold aspects of marine ecosystem modifications have been largely discussed, including the distinction between fluctuations and the role of episodic events in coastal community variations (see, e.g., [5,6]).

Along with ongoing environmental degradation from a local scale to the global scale, accepted thresholds for environmental conditions can be continually lowered, affecting the perception and awareness of community changes. Lacking past information or experience, each new generation of ecologists might accept their rising situation as the norm, a sociopsychological phenomenon known as the shifting baseline syndrome (SBS) [7].

Global changes of biota in marine environments are nowadays of emerging relevance because of the acceleration induced in recent decades by an increase in anthropogenic pressure (i.e., pollution, coastal constructions, overfishing), particularly in the Mediterranean Sea, whose enclosed basin magnifies global warming effects on water temperature [8–10].

Accordingly, instead of exhaustively reviewing changes in ecological studies, we focus here on long-term (i.e., from decades up to a century) relevant changes in marine biodiversity of Mediterranean benthic communities along the Italian coast.

We report six different cases of perception and detection of variations, which can be assessed due to the availability of historical records that provide clear baselines to understand the possible ecosystem change paths.

We speculate on the importance of these historical data as reservoirs of ecological memory, allowing the understanding of current and future changes. Moreover, we stress the pivotal role of taxonomists, who should be considered an essential link between old and new generations of ecologists. This, in turn, could be basal for preserving the knowledge required to understand changes in the benthic communities over time, as well as to allow the new generations of ecologists to avoid SBS.

2. Study Cases

2.1. Ficopomatus Reef

The serpulid polychaete *Ficopomatus enigmaticus* (Fauvel, 1923) had spread into the Mediterranean Sea since the beginning of the last century through unaided dispersal from native regional borders [11–13]. The first historical record of its massive tube agglomerations along the Italian coast dates back to 1919 [14]. This ecosystem engineer edifies conspicuous reefs in brackish water systems, consisting of complex clumps of cemented calcareous tubes (Figure 1a–d) that offer refuge, food, and habitat for reproduction to many other benthic organisms. Consequently, its presence strongly modifies the distribution and increases the abundance and diversity of brackish Mediterranean benthic fauna [15–22]. *Ficopomatus* reefs grow quickly and spectacularly in coastal lagoons, which are progressively filled up by reefs and skeletal debris of such serpulid and the associated benthic organisms that may change the ecosystem dynamics. For decades, the Italian *Ficopomatus* reefs have remained comparatively stable, and they are considered a characteristic habitat of the eurythermal and euryhaline lagoons [23].

Figure 1. *Ficopomatus* reef at Fiumicino, Rome (Italy): calcareous tubes cemented one to another (**a**); details of the bioconstruction, forming belts that fringe the shoreline in a continuous layer up to 0.5 m thick, showed at progressively increased distances (**b–d**). Photo credit: A. Bonifazi.

2.2. Sabellaria Reef

Sabellaria alveolata (Linnaeus, 1767), a gregarious sabellariid polychaete, is able to build compact bioconstructions in the intertidal and shallow subtidal (Figure 2). They occur on sandy or hard bottoms in both the Northeastern Atlantic and the Mediterranean. Their massive bioconstructions (i.e., sheets, hummocks, banks) of cemented tube aggregates strongly modify coastal marine habitats as they support a high diversity and act as natural barriers against coastal erosion [24,25]. Along the Italian coasts, large *S. alveolata* reefs occur in Latium [24,26–28] and Sicily [29], while the first remarkable reef made by the cogeneric *Sabellaria spinulosa* (Leuckart, 1849) was recently reported along the Apulian coast [30,31]. The Latium and Sicily *S. alveolata* reefs have been well known since the 1950s [32–34] when its "pristine" condition represented a sound ecological baseline. Differences in reef structure and morphology mainly result from combining the current developmental phase with environmental conditions, balanced by the destruction/construction cycle [24,25]. Therefore, assessing long-term changes relies upon the comparison of present and past status; in the case of both Latium and Sicily reefs, there have been no noticeable changes [24,25,29,35], so this demonstrates that they have been thriving for over half a century.

Figure 2. *Sabellaria alveolata* reef at Tor Caldara, Tyrrhenian Latium coast, Italy, in the upper infralittoral (**a–c**) and emerging during low tide (**d,e**); detail of the peculiar honeycomb-like bioconstruction, showing the tube openings with the "sand crown" as a diagnostic character (**f**). Photo credit: A. Bonifazi and D. Ventura.

2.3. Posidonia oceanica Meadow

The Mediterranean endemic seagrass *Posidonia oceanica* (Linnaeus) Delile, 1813 is characterized by its high leaf shoots thriving in the water column and its matte setting on the seabed. It is the main bioengineer species along Mediterranean coastal areas (Figure 3), where it creates an original and productive ecosystem that strongly promotes benthic and fish communities, supports marine biodiversity, and furnishes several ecosystem services [36–38]. Due to its ecological, heritage, and economic relevance, the meadows are identified as a priority habitat (1120 *Posidonia oceanica* bed) for conservation and are included in the Natura 2000 marine sites (Habitat Directive 92/43/CEE) and in the Barcelona Convention (16.02.1976) while being protected by different European directives (WFD, 2000/60/EC; MSFD, 2008/56/EC) and national laws. Nevertheless, many meadows have suffered a decline caused by anthropogenic disturbances altering their spatial extent and density [39–44]. Consequently, there has been strong concern and warnings about their conservation status, leading to extensive investigations and monitoring that have revealed their main change trajectories during the last decades [45,46]. The large size of the plant and the complex structure of the meadows facilitate the understanding of the time and causes of damage; virtually all of them are attributable to anthropogenic activities in the coastal areas. In the Middle Tyrrhenian Sea, along the Latium coasts, the large body of historical data available has allowed an effective comparison with the present meadow conditions, including the magnitudes of impact and the main causes of its decline [27,39,41,42,46–51], information of crucial interest for conservation and management strategies in the region.

Figure 3. *Posidonia oceanica* meadow at Isola del Giglio (Tuscan Archipelago, Tyrrhenian Sea, Italy) in good ecological status (**a**,**b**); *P. oceanica* flowering (**c**); example of an impacted meadow, showing the lower regression limit with dead matte and dead shells of *Pinna nobilis* Linnaeus, 1758 (**d**). Photo credit: D. Ventura.

2.4. "Sponge Garden" of La Strea Bay

The "sponge garden" of La Strea Bay, along the Ionian coast of Apulia (Italy), was a unique ecosystem dominated by an extraordinarily diverse number of sponge species of different sizes. Particularly, *Geodia cydonium* (Linnaeus, 1767; Figure 4), an Atlantic–Mediterranean sponge commonly living in sheltered coastal waters, thrived as a dense population in the bay, composed of specimens that were variable in dimensions, reaching up to 40–100 cm in diameter [52–54]. Both sessile and nonsessile specimens coexisted, the latter being able to roll on a soft bottom, dragged by slow circular currents [52,53]. Since 1976, *G. cydonium* has been considered an "umbrella species" for the entire sponge garden, where it has played a crucial role in harboring invertebrates and algae, offering sites for fish spawning and nurseries and strongly contributing to increased biodiversity in La Strea Bay [54–56]. La Strea Bay was not included in the Marine Protected Area of Porto Cesareo, established in 1997 [57]. As a consequence, a dock for mooring recreational boats was built, leading to an abrupt change in the benthic communities, including the loss of the "sponge garden" (G. Corriero personal communication). The comparison between present and past pristine conditions has provided pivotal information on the last change of the sponge assemblage.

Figure 4. La Strea Bay at the Ionian Apulian coast, Lecce (Italy) showing the dock for the recreational boat mooring (**a**,**b**), photo credit: M.F. Gravina; massive *Geodia cydonium* subspherical specimen (**c**); *Geodia cydonium* specimen covered by epibionts, which protect this sciaphilous sponge from high solar radiation (**d**). Photo by courtesy of G. Corriero, University of Bari, Italy.

2.5. Introduction of Ruditapes philippinarum

The human-mediated changes caused by the Manila clam *Ruditapes philippinarum* (A. Adams and Reeve, 1850) (Figure 5) in sandy seabeds are well known and clearly documented. The species was introduced in Italian North Adriatic brackish waters through an aquaculture program in 1983 [58,59], resulting in a quick adaptation due to its great resistance and fast growth. Three years later, it was breeding freely and had colonized all suitable areas, where it completely replaced the carpet-shell clam native *Ruditapes decussatus* (Linnaeus, 1758). The rapid growth and high densities of the exotic bivalve caused abrupt changes in harvesting technology and the fishery market. However, the local soft-bottom assemblages did not show remarkable changes in biodiversity [58]. The only remaining populations of *R. decussatus* occur in some brackish lagoons and ponds along the Sardinian coast of the Tyrrhenian Sea (e.g., Tortolì, San Giovanni, Merceddì–Corru s'ittiri, Santa Gilla, Calich) and Latium (Lago di Paola) [60–62], where they are extensively cultured and collected for human consumption. The definite date and site of the introduction constitute a clear before-and-after impact example, allowing us to suggest specific timing and methods for conservation plans of the still-surviving populations of the autochthonous *R. decussatus*.

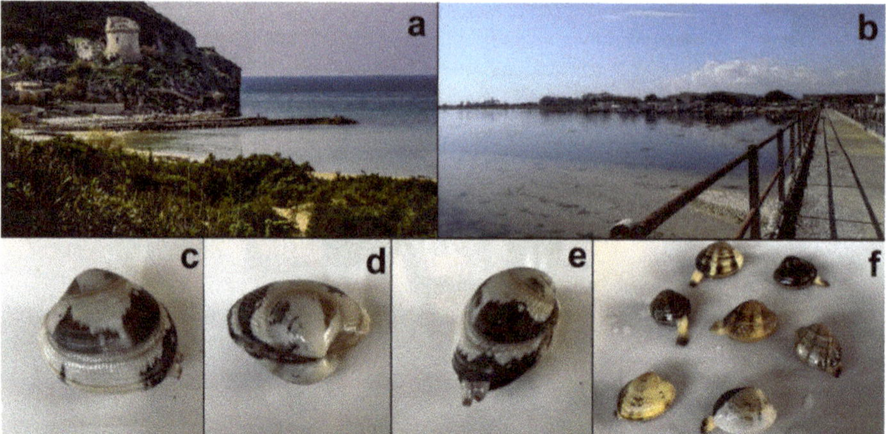

Figure 5. Lago di Paola, Latium, Italy: detail of the mouth channel, photo credit: D. Ventura (**a**) and Santa Gilla lagoon, Sardinia, Italy, photo by courtesy of Serenella Cabiddu, University of Cagliari, Italy (**b**) as examples of sites where populations of the autochthonous *Ruditapes decussatus* occur; specimens of the allochthonous *Ruditapes philippinarum,* showing shells and the almost-fused siphons as diagnostic characters (**c**–**f**). Photo credit: A. Bonifazi.

2.6. Fouling Community of the Mar Grande of Taranto

The fouling community of the Mar Grande of Taranto has been investigated, and its distinctive nature was exhibited from 1969 up to the next thirty-five years [63–66]. The arrival of nonindigenous species during the last fifteen years significantly changed the structure and function of the entire community [67–74]. Particularly, two allochthonous sabellid polychaetes, *Branchiomma luctuosum* (Grube, 1870) and *Branchiomma boholense* (Grube, 1878), appeared. At first, *B. luctuosum* was highly invasive and outcompeted the dominant native fan worm *Sabella spallanzanii* (Gmelin, 1791; Figure 6a,b) [75]. Then, *B. boholense* spread and became dominant, together with *B. luctuosum* (Figure 6c) [76]. Nowadays, the fouling assemblage is highly diverse and includes all three sabellids, although *S. spallanzanii* is still the most abundant (Figure 6d) [77]. The present vs. past community comparison highlights changes that are still ongoing.

Figure 6. Fouling at the Mar Grande of Taranto, Ionian Sea, Italy: the autochthonous fan worm *Sabella spallanzanii*, the only dominant species thirty years ago (**a**); the allochthonous *Branchiomma luctuosum* (**b**); the allochthonous *B. luctuosum* with *B. boholense*, invasive for fifteen years (**c**); *Sabella spallanzanii*, the most abundant species nowadays (**d**). Photo credit: M. Del Pasqua, A. Giangrande and F. Mastrototaro.

3. Perception of Changes

Perception means to become aware of reality, consequently implementing a cognitive process. In a biodiversity context, change perception is a basic condition to assess variations, detect reliable causes, and foresee possible consequences, depending on the type of environmental impact and anthropic alteration. In our study cases, the evaluation of change perception in established benthic communities depended on different criteria (Table 1). It also required a comparison of different spatial and temporal status and the identification of the main causal processes and factors. A temporal baseline, representing the boundary between pristine and impacted conditions, is the primary step to check changes in a diachronic sequence (Table 1: Temporal baseline reference). "Pristine status" is generally considered the natural, original ecosystem condition, i.e., the "good status" to be expected, while an "impacted status" has suffered human pressures and is, thus, deteriorated and "poor". Communities generally show high diversity in the former and low diversity in the latter. Assessing community status is particularly relevant in conservation science to both protect species, habitats, and ecosystems and to ward biodiversity from being excessively eroded. Moreover, the analysis of time changes allows us to infer possible consequences on ecosystem functioning and biodiversity, raising its interest beyond the scope of ecologists.

28

Table 1. Criterions of understanding changes of the study cases. In brackets the references of the temporal baseline.

Study Cases	Historical Memory			Conservation of Taxonomic and Ecological Data			Popularity of Dominant Species	Change Perception
	Temporal Baseline Reference	Available Literature	Years of Investigation	Museum Collections	Scientific Collections	Available Images (Maps, Charts)		
Ficopomatus reefs	1919 [14]	Lindegg, 1934; Fauvel, 1938; Rullier, 1955; Peres & Picard, 1964; Gravina et al., 1989; Bianchi & Morri, 1996; Bianchi e Morri, 2001; Nonnis Marzano et al., 2003; Cardone et al., 2013; Giangrande & Gravina, 2015	60	no	yes	yes	yes	no change
Tor Caldara *Sabellaria* reef	1956 [32]	Taramelli Rivosecchi, 1961; La Porta & Nicoletti, 2009; Ventura et al., 2018; 2020; Bonifazi et al., 2019; Lisco et al., 2020	60	no	yes	yes	yes	no change
Posidonia meadow, Middle Tyrrhenian Sea-Latium coast	1959–1961 [47,46]	Ardizzone & Migliuolo, 1982; Ardizzone & Pelusi, 1984; Ardizzone & Belluscio, 1996; Diviacco et al., 2001; Ardizzone et al., 2006; Telesca et al., 2015; Ventura et al., 2017; 2018; 2020	40	no	no	yes	yes	in progress
La Strea Sponge garden	1976 [53]	Corriero et al., 1984; Gherardi et al., 2001; Corriero et al., 2004; Mercurio et al., 2006; 2007	40	yes	yes	yes	yes	completed
Ruditapes philippinarum, North Adriatic Sea	1983 [58]	Breber, 1985; 2002	40	no	no	no	yes	completed
Fouling of Mar Grande of Taranto	1969 [63]	Gherardi & Lepore, 1974; Tursi et al., 1976; 1982; Giangrande et al., 2000; Brunetti & Mastrototaro, 2004; Mastrototaro & Brunetti, 2006; Longo et al., 2007; Pierri et al., 2010; Petrocelli et al., 2013; Giangrande et al., 2014, Del Pasqua et al., 2018; Lezzi et al., 2018a; 2018b	50	yes	yes	yes	no	in progress

Assessing variations during several decades to a century requires repeated monitoring to understand change trends and rates [78–80]. This has led us to search for the main available studies and their extent in our study cases (Table 1: Available literature; Years of investigation) as the core of "ecological memory", allowing the past sequences of community events to be encoded and stored. This way, the variations in assemblage structure, dynamics, and biodiversity remain as ecological memories, allowing us to understand their current and future structures and functioning [78,81]. Ecological memories concerning benthic communities are also stored in specimen collections and distribution maps/charts in both museums and academic offices (Table 1: Museum and scientific collections; Available images). Taxonomic and ecological references are not simply inventories but reservoirs of historical community components, allowing present vs. past pattern comparisons. Museum and scientific collections provide key relevant cues to studying ongoing and predicting future community changes. In turn, the large gaps typically suffered by most historical quantitative datasets can be bridged by descriptive observations and qualitative datasets that specifically regard the emblematic and common species (Table 1: Popularity of species).

In our study cases, the available literature and images were the main source of historical memory of the *Ficopomatus* reefs, as well as the popularity of the species, which is widespread among the lagoon fishermen. The existence of images constitutes the cartographic baseline support, allowing us to perceive the changes in the *Posidonia* meadows. Taxonomic skills were most relevant in the fouling of the Mar Grande of Taranto and the *Ficopomatus* reefs, as distinguishing between different species is required to assess whether defining their biological traits allows the perception of changes in community structure and interspecific interactions between species. Beyond the other memory reservoirs, museum collections are particularly relevant in the case of the La Strea sponge garden. Indeed, the specimens preserved in the Museum of Porto Cesareo were key in evaluating size variations of *G. cydonium* in the La Strea sponge garden, as well as assessing its population dynamics. The popularity in the case of the *Sabellaria* reefs of Tor Caldara also played a key role, as the status of the local reefs was well known by several human generations because they occur in recreational bath areas. Similarly, the popularity and commercial interest of *R. philippinarum* have been very useful to perceive changes.

4. Taxonomy and Comprehension of Change

Morphological and functional knowledge of species is key to understanding variations and interpreting changes; this is the main purpose of taxonomy. Thus, this science branch is certainly crucial to understanding causes and detecting trends and rates of change in ecosystems over time. Long-term studies require periods often exceeding individual professional lifetimes. Taxonomic works are long-lasting and require specific skills that must be handed over from one generation to the next to preserve historical knowledge and allow new generations to have a more comprehensive awareness of biological changes, as well as to prevent them from suffering SBS [82,83]. The gradual change in human perception of environmental conditions often results in increasing tolerance to environmental degradation in parallel with increasing ignorance of past conditions, leading to harmful consequences on environmental and cultural heritage and conservation (see, e.g., [7,81,84,85]). Taxonomy is, thus, a basic science to deal with changes in biodiversity and ecosystems. Taxonomists also record the background on species in museum collections, which are rich in informative contents and baselines [86–91]. Museum collections have also been used in research on marine biodiversity changes, with particular regard to depauperation/loss of species (see, e.g., [92–94] for tropical and [95] for Italian species, together with examples of personal observations (M.F. Gravina) in the Civic Museum of Rome regarding the popular Mediterranean monk seal *Monachus monachus* (Hermann, 1779) and the sea lamprey *Petromyzon marinus* Linnaeus, 1758, from the Sardinian and Latium coasts, respectively. Unfortunately, collections are too often overlooked because the taxonomic expertise that allows the interpretation of the information associated with the specimens deposited is disappearing together with the specialists.

The identification of organisms to the species level is pivotal in the study of biodiversity. It appears to be a conflicting and obvious paradox that in the present era of biodiversity, as endorsed since the Rio Convention on Biological Diversity, taxonomy based on phenotypes still remains a marginal science [96–98], notwithstanding its crucial role in understanding causes and processes of changing ecosystems over time (see, e.g., [99]). It is really nonsense to prioritize molecular approaches rather than phenotypical studies on species because they both encompass key aspects of the same organism! Moreover, changes in diversity and community structure are particularly studied in monitoring and environmental quality assessment programs, which addressed possible disturbance causes and anthropic impacts according to the requirements of National and European current legislation (WFD, 2000/60/EC; MSFD, 2008/56/EC). In such studies, the identification of organisms at the species level is a key requisite. In other words, the survey of biodiversity changes requires good taxonomic work!

With this in mind, we hope for an increase in the taxonomic workforce through retraining of taxonomic schools to preserve the knowledge of older generations while attracting new generations to taxonomy, allowing them to have a more comprehensive awareness of biological changes in our changing Mediterranean biota.

5. Conclusive Remarks

The large number of studies on synchronic and diachronic ecosystem dynamics conducted in the Mediterranean has stressed different types of changes in community structures and diversity patterns as unquestionable evidence derived from variation, one of the most characteristic traits of the history of life.

Our study cases have highlighted the importance of having taxonomic and ecological knowledge of benthic communities, coupled with the correct interpretation of the available species lists, as a reference baseline to evaluate the variations occurring in the present and assess long-term changes in the benthic communities. Knowing the species has, thus, a pivotal role in understanding variations and interpreting changes. Taxonomists know the morphological, molecular, and ecological traits of the species and are essential to understanding changes in biodiversity. This certainly includes the extensive but often overlooked knowledge on the species that is presented by museum collections, which need to be revalued as reference baselines, reliably supporting observational and literature data. In addition, new collections need to be set up to be compared, a particularly useful tool during identification procedures. In this way, taxonomists could provide robust identifications.

We conclude that the better we know the past ecosystem composition, the more we will understand the present ecosystem functioning and the better we will be able to foresee its future. In the era of biodiversity, we support reevaluating the study of species as main ecosystem actors. Therefore, we strongly recommend new investments in taxonomic schools to assure intergenerational connectivity, together with the transmission of cultural and environmental heritage, a key aspect to understanding and managing our changing ecosystems.

Author Contributions: Conceptualization, M.F.G. and A.G.; methodology, M.F.G., A.G., A.B., M.D.P., J.G., M.L. and D.V.; formal analysis, M.F.G. and A.G.; data curation, M.F.G. and A.G.; methodology, M.F.G., A.G., A.B., M.D.P., J.G., M.L. and D.V.; writing—original draft preparation, M.F.G., A.G., A.B., M.D.P., J.G., M.L. and D.V.; writing—review and editing, M.F.G., A.G., A.B., M.D.P., J.G., M.L. and D.V.; supervision, D.V. All authors have read and agreed to the published version of the manuscript.

Funding: This study was partially (study case of the fouling of the Mar Grande of Taranto) supported by the project "Remedialife" (LIFE16 ENV/IT/000343) funded by the European Commission.

Conflicts of Interest: The authors declare no conflict of interest.

References

1. Sarà, M. Persistence and changes in marine benthic communities. *Nova Thalassia* **1985**, *7*, 7–30.
2. Sarà, M. Cambiamenti ed evoluzione negli ecosistemi marini. In *21° Seminario Sulla Evoluzione Biologica e i Grandi Problemi Della Biologia. Evoluzione Degli Ecosistemi*; Accademia Nazionale dei Lincei: Roma, Italy, 1995; Volume 1, pp. 17–30.
3. Bianchi, C.N.; Boero, F.; Fonda Umani, S.; Morri, C.; Vacchi, M. Successione e cambiamento negli ecosistemi marini. *Biol. Mar. Mediterr.* **1998**, *5*, 117–135.
4. Carson, H.L. The genetics of speciation at the diploid level. *Am. Nat.* **1975**, *109*, 83–92. [CrossRef]
5. Boero, F. Fluctuations and variations in coastal marine environments. *Mar. Ecol.* **1994**, *15*, ventura3–ventura25. [CrossRef]
6. Boero, F. Episodic events: Their relevance to ecology and evolution. *Mar. Ecol.* **1996**, *17*, 237–250. [CrossRef]
7. Soga, M.; Gaston, K.J. Shifting baseline syndrome: Causes, consequences, and implications. *Front. Ecol. Envol.* **2018**, *16*, 222–230. [CrossRef]
8. Ulbrich, U.; May, W.; Li, L.; Lionello, P.; Pinto, J.G.; Somot, S. The Mediterranean climate change under global warming. In *Developments in Earth and Environmental Sciences*; Lionello, P., Malanotte Rizzoli, P., Boscolo, R., Eds.; Elsevier: Amsterdam, The Netherlands, 2006; Volume 4, pp. 399–415.
9. Rivetti, I.; Fraschetti, S.; Lionello, P.; Zambianchi, E.; Boero, F. Global warming and mass mortalities of benthic invertebrates in the Mediterranean Sea. *PLoS ONE* **2014**, *9*, e115655. [CrossRef]
10. Bianchi, C.N.; Azzola, A.; Parravicini, V.; Peirano, A.; Morri, C.; Montefalcone, M. Abrupt change in a subtidal rocky reef community coincided with a rapid acceleration of sea water warming. *Diversity* **2019**, *11*, 215. [CrossRef]
11. Fauvel, P. Annelida polychaeta della Laguna di Venezia. *Mem. R. Com. Talass. Ital.* **1938**, *246*, 1–27.
12. Rullier, F. Quelques stations nouvelles de *Mercierella enigmatica* Fauvel sur le littoral méditerranéen, aux environs de Marseille et sur la côte italienne. *Vie Milieu* **1955**, *6*, 74–82.
13. Servello, G.; Andaloro, F.; Azzurro, E.; Castriota, L.; Catra, M.; Chiarore, A.; Crocetta, F.; D'Alessandro, M.; Denitto, F.; Froglia, C.; et al. Marine alien species in Italy: A contribution to the implementation of descriptor D2 of the marine strategy framework directive. *Mediterr. Mar. Sci.* **2019**, *20*, 1–48. [CrossRef]
14. Lindegg, G. La "*Mercierella enigmatica*" Fauvel nello stagno di Cabras in Sardegna. *Natura* **1934**, *25*, 135–145.
15. Tenerelli, V. Sulla presenza di *Mercierella enigmatica* Fauvel lungo la costa orientale di Sicilia (Polychaeta, Serpulidae). *Boll. Zool.* **1966**, *24*, 735–748.
16. Gravina, M.F.; Ardizzone, G.D.; Scaletta, F.; Chimenz, C. Descriptive analysis and classification of benthic communities in some Mediterranean lagoons (Central Italy). *Mar. Ecol.* **1989**, *10*, 141–166. [CrossRef]
17. Bianchi, C.N.; Morri, C. *Ficopomatus* 'Reefs' in the Po river delta (Northern Adriatic): Their constructional dynamics, biology and influences on the brackish-water biota. *Mar. Ecol.* **1996**, *17*, 51–66. [CrossRef]
18. Bianchi, C.N.; Morri, C. The battle is not to the strong: Serpulid reefs in the lagoon of Orbetello (Tuscany, Italy). *Estuar. Coast. Shelf Sci.* **2001**, *53*, 215–220. [CrossRef]
19. Nonnis Marzano, C.; Scalera Liaci, L.; Fianchini, A.; Gravina, M.F.; Mercurio, M.; Corriero, G. Distribution, persistence and change in macrobenthos of the lagoon of Lesina (Apulia, southern Adriatic Sea). *Oceanol. Acta* **2003**, *26*, 57–66. [CrossRef]
20. Nonnis Marzano, C.; Baldacconi, R.; Fianchini, A.; Gravina, M.F.; Corriero, G. Settlement seasonality and temporal changes in hard substrate macrozoobenthic communities of Lesina lagoon (Apulia, Southern Adriatic Sea). *Chem. Ecol.* **2007**, *23*, 479–491. [CrossRef]
21. Cardone, F.; Corriero, G.; Fianchini, A.; Gravina, M.F.; Nonnis Marzano, C. Biodiversity of transitional waters: Species composition and comparative analysis of hard bottom communities from the South-Eastern Italy coast. *J. Mar. Biol. Assoc. UK* **2013**, *94*, 25–34. [CrossRef]
22. Giangrande, A.; Gravina, M.F. Brackish-water polychaetes, good descriptors of environmental changes in space and time. *Transit. Water Bull.* **2015**, *9*, 42–55.
23. Pérès, J.M.; Picard, J. Nouveau manuel de bionomie benthique de la Méditeraneée. *Recl. Trav. Stn. Mar. Endoume* **1964**, *31*, 1–37.
24. Bonifazi, A.; Lezzi, M.; Ventura, D.; Lisco, S.; Cardone, F.; Gravina, M.F. Macrofaunal biodiversity associated with different developmental phases of a threatened Mediterranean *Sabellaria alveolata* (Linnaeus, 1767) reef. *Mar. Environ. Res.* **2019**, *145*, 97–111. [CrossRef] [PubMed]

25. Lisco, S.N.; Acquafredda, P.; Gallicchio, S.; Sabato, L.; Bonifazi, A.; Cardone, F.; Corriero, G.; Gravina, M.F.; Pierri, C.; Moretti, M. The sedimentary dynamics of *Sabellaria alveolata* bioconstructions (Ostia, Tyrrhenian Sea, central Italy). *J. Palaeogeogr.* **2020**, *9*, 1–18. [CrossRef]

26. La Porta, B.; Nicoletti, L. *Sabellaria alveolata* (Linnaeus) reefs in the central TyrrhenianSea (Italy) and associated polychaete fauna. *Zoosymposia* **2009**, *2*, 527–536. [CrossRef]

27. Ventura, D.; Bonifazi, A.; Gravina, M.F.; Belluscio, A.; Ardizzone, G. Mapping and classification of ecologically sensitive marine habitats using unmanned aerial vehicle (UAV) imagery and object-based image analysis (OBIA). *Remote Sens.* **2018**, *10*, 1331. [CrossRef]

28. Ventura, D.; Dubois, S.F.; Bonifazi, A.; Jona Lasinio, G.; Seminara, M.; Gravina, M.F.; Ardizzone, G.D. Integration of close-range underwater photogrammetry with inspection and mesh processing software: A novel approach for quantifying ecological dynamics of temperate biogenic reefs. *Remote Sens. Ecol. Conserv.* **2020**. [CrossRef]

29. Schimmenti, E.; Musco, L.; Lo Brutto, S.; Mikac, B.; Nygren, A.; Badalamenti, F. Mediterranean record of *Eulalia ornata* (Annelida: Phyllodocidae) corroborating its fidelity link with the *Sabellaria alveolata* reef habitat. *Medit. Mar. Sci.* **2016**, *17*, 359–370. [CrossRef]

30. Lisco, S.N.; Moretti, M.; Moretti, V.; Cardone, F.; Corriero, G.; Longo, C. Sedimentological features of *Sabellaria spinulosa* biocontructions. *Mar. Pet. Geol.* **2017**, *87*, 203–212. [CrossRef]

31. Gravina, M.F.; Cardone, F.; Bonifazi, A.; Bertrandino, M.S.; Chimienti, G.; Longo, C.; Nonnis Marzano, C.; Moretti, M.; Lisco, S.; Moretti, V.; et al. *Sabellaria spinulosa* (Polychaeta, Annelida) reefs in the Mediterranean Sea: Habitat mapping, dynamics and associated fauna for conservation management. *Estuar. Coast. Shelf Sci.* **2018**, *200*, 248–257. [CrossRef]

32. Giordani Soika, A. Scogliera pseudocorallina intercotidale di *Sabellaria alveolata* (L.) nelle coste del Lazio (Ann. Polych.). *Boll. Mus. Civico Storia Nat. Venezia.* **1956**, *9*, 11–13.

33. Taramelli Rivorecchi, E. Osservazioni sulle biocenosi del banco a Sabellaria di Lavinio. *Rend. Accad. Naz. XL* **1961**, *12*, 147–157.

34. Molinier, R.; Picard, J. Notes biologiques à propos d'un voyage d'étude sur les côtes de Sicilie. *Ann. Inst. Oceanogr.* **1953**, *28*, 163–188.

35. Ingrosso, G.; Abbiati, M.; Badalamenti, F.; Bavestrello, G.; Belmonte, G.; Cannas, R.; Benedetti-Cecchi, L.; Bertolino, M.; Bevilacqua, S.; Bianchi, C.N.; et al. Mediterranean bioconstructions along the Italian coast. *Adv. Mar. Biol.* **2018**, *79*, 61–136. [PubMed]

36. Boudouresque, C.F.; Mayot, N.; Pergent, G. The outstanding traits of the functioning of the *Posidonia oceanica* seagrass ecosystem. *Biol. Mar. Medit.* **2006**, *13*, 109–113.

37. Vassallo, P.; Paoli, C.; Rovere, A.; Montefalcone, M.; Morri, C.; Bianchi, C.N. The value of the seagrass *Posidonia oceanica*: A natural capital assessment. *Mar. Poll. Bull.* **2013**, *75*, 157–167. [CrossRef]

38. Boudouresque, C.F.; Pergent, G.; Pergent-Martini, C.; Ruitton, S.; Thibaut, T.; Verlaque, M. The necromass of the *Posidonia oceanica* seagrass meadow: Fate, role, ecosystem services and vulnerability. *Hydrobiologia* **2016**, *781*, 25–42. [CrossRef]

39. Ardizzone, G.D.; Belluscio, A. Le praterie di *Posidonia oceanica* delle coste laziali. In *Il Mare del Lazio*; Università degli Studi di Roma "La Sapienza": Roma, Italy, 1996; pp. 194–217.

40. Diviacco, G.; Spada, E.; Virno Lamberti, C. *Le fanerogame marine del Lazio*; Istituto Centrale per la Ricerca Scientifica e Tecnologica Applicata al Mare: Roma, Italy, 2001; p. 113.

41. Ardizzone, G.D.; Belluscio, A.; Maiorano, L. Long-term change in the structure of a *Posidonia oceanica* landscape and its reference for a monitoring plan. *Mar. Ecol.* **2006**, *27*, 299–309. [CrossRef]

42. Boudouresque, C.F.; Bernard, G.; Pergent, G.; Shili, A.; Verlaque, M. Regression of Mediterranean seagrasses caused by natural processes and anthropogenic disturbances and stress: A critical review. *Bot. Mar.* **2009**, *52*, 395–418. [CrossRef]

43. Montefalcone, M.; Albertelli, G.; Morri, C.; Parravicini, V.; Bianchi, C.N. Legal protection is not enough: *Posidonia oceanica* meadows in marine protected areas are not healthier than those in unprotected areas of the northwest Mediterranean Sea. *Mar. Poll. Bull.* **2009**, *58*, 515–519. [CrossRef]

44. Vacchi, M.; De Falco, G.; Simeone, S.; Montefalcone, M.; Morri, C.; Ferrari, M.; Bianchi, C.N. Biogeomorphology of the Mediterranean *Posidonia oceanica* seagrass meadows. *Earth Surf. Process. Landf.* **2017**, *42*, 42–54. [CrossRef]

45. Alami, S.; Bonacorsi, M.; Clabaut, P.; Jouet, G.; Pergent-Martini, C.; Pergent, G.; Sterckeman, A. Assessment and quantification of the anthropic impact on the *Posidonia oceanica* seagrass meadow. In *5th Mediterranean Symposium on Marine Vegetation*; Langar, H., Bouafif, C., Ouerghi, A., Eds.; RAC/SPA Publ.: Tunis, Tunisia, 2014; pp. 34–39.

46. Telesca, L.; Belluscio, A.; Criscoli, A.; Ardizzone, G.D.; Apostolaki, E.T.; Fraschetti, S.; Gristina, M.; Knittweis, L.; Martin, C.S.; Pergent, G.; et al. Seagrass meadows (*Posidonia oceanica*) distribution and trajectories of change. *Sci. Rep.* **2015**, *5*, 12505. [CrossRef] [PubMed]

47. Fusco, N. *Dal promontorio dell'Argentario a Fiumicino*; Direzione Generale Pesca Marittima: Roma, Italy, 1959.

48. Fusco, N. *Da Capo Circeo a Capo Miseno*; Direzione Generale Pesca Marittima: Roma, Italy, 1961.

49. Ardizzone, G.D.; Migliuolo, A. Modificazioni di una prateria di *Posidonia oceanica* (L.) Delile del Medio Tirreno sottoposta ad attività di pesca a strascico. *Nat. Sicil.* **1982**, *IV* (Suppl. VI), 509–515.

50. Ardizzone, G.D.; Pelusi, P. Yield and damage evaluation of bottom trawling on Posidonia meadows. In *First International Workshop on Posidonia oceanica Beds*; Boudouresque, C.F., Jeudy de Grissac, A., Oliver, J., Eds.; GIS Posidonie Publications: Porquerolles, France, 1984; Volume 1, pp. 63–67.

51. Ventura, D.; Bonifazi, A.; Gravina, M.F.; Ardizzone, G.D. Unmanned aerial systems (UASs) for environmental monitoring: A review with applications in coastal habitats. In *Aerial Robots-Aerodynamics, Control and Applications*; Lopez Mejia, L.D., Ed.; InTech Open: London, UK, 2017; pp. 165–184. Available online: https://www.intechopen.com/books/aerial-robots-aerodynamics-control-and-applications (accessed on 1 September 2020).

52. Mercurio, M.; Corriero, G.; Gaino, E. A 3-year investigation of sexual reproduction in *Geodia cydonium* (Jameson 1811) (Porifera, Demospongiae) from a semi-enclosed Mediterranean bay. *Mar. Biol.* **2007**, *151*, 1491–1500. [CrossRef]

53. Parenzan, P. Un habitat marino di tipo subtropicale a Porto Cesareo. In *Atti del VI Simposio Nazionale per la Conservazione Della Natura*; Scalera Liaci, L., Ed.; Cacucci Editore: Bari, Italy, 1976; pp. 151–157.

54. Corriero, G.; Pansini, M.; Sarà, M. Sui poriferi della insenatura della Strea a Porto Cesareo (Lecce). *Thalass. Salentina* **1984**, *14*, 3–10.

55. Gherardi, M.; Giangrande, A.; Corriero, G. Epibiontic and endobiontic polychaetes of *Geodia cydonium* (Porifera, Demospongiae) from the Mediterranean Sea. *Hydrobiologia* **2001**, *443*, 87–101. [CrossRef]

56. Mercurio, M.; Longo, C.; Corriero, G. Modificazioni della fauna a Poriferi nella insenatura della Strea di Porto Cesareo (Mar Ionio). *Biol. Mar. Medit.* **2006**, *13*, 257–260.

57. Corriero, G.; Gherardi, M.; Giangrande, A.; Longo, C.; Mercurio, M.; Musco, L.; Nonnis Marzano, C. Inventory and distribution of hard bottom fauna from the marine protected area of Porto Cesareo (Ionian Sea): Porifera and Polychaeta. *Ital. J. Zool.* **2004**, *71*, 237–245. [CrossRef]

58. Breber, P. L'introduzione e l'allevamento in Italia dell'Arsella del Pacifico, *Tapes semidecussatus* Reeve (Bivalvia: Veneridae). *Oebalia* **1985**, *11*, 675–680.

59. Breber, P. Introduction and acclimatisation of the Pacific carpet clam *Tapes philippinarum*, to Italian waters. In *Invasive Aquatic Species of Europe. Distribution, Impacts and Management*; Leppäkoski, E., Gollasch, S., Olenin, S., Eds.; Springer: Rotterdam, The Netherlands, 2002; pp. 120–126.

60. Chessa, L.A.; Paesanti, F.; Pais, A.; Scardi, M.; Serra, S.; Vitale, L. Perspectives for development of low impact aquaculture in a Western Mediterranean lagoon: The case of the carpet clam Tapes decussatus. *Aquac. Int.* **2005**, *13*, 147–155. [CrossRef]

61. Pais, A.; Chessa, L.A.; Serra, S.; Ruiu, A. An alternative suspended culture method for the Mediterranean carpet clam, *Tapes decussatus* (L.), in the Calich lagoon (North Western Sardinia). *Biol. Mar. Medit.* **2006**, *13*, 134–135.

62. Mura, L.; Cossu, P.; Cannas, A.; Scarpa, F.; Sanna, D.; Dedola, G.L.; Floris, R.; Lai, T.; Cristo, B.; Curini-Galletti, M.; et al. Genetic variability in the Sardinian population of the manila clam, *Ruditapes philippinarum*. *Biochem. Syst. Ecol.* **2012**, *41*, 74–82. [CrossRef]

63. Parenzan, P. Il Mar Piccolo e il Mar Grande di Taranto. *Thalass. Salentina* **1969**, *3*, 19–36.

64. Gherardi, M.; Lepore, E. Insediamenti stagionali delle popolazioni fouling del mar Piccolo di Taranto. In *Atti IV Simposio Nazionale sulla Conservazione della Natura*; Università di Bari: Bari, Italy, 1974; pp. 235–258.

65. Tursi, A.; Gherardi, M.; Lepore, E.; Chieppa, M. Settlement and growth of Ascidians on experimental panels in two harbours of Southern Italy. In Proceedings of the IV International Congress on Marine Corrosion and Fouling, Antibes, Juan-les-Pins, France, 14–18 June 1976; pp. 535–543.

66. Tursi, A.; Matarrese, A.; Sciscioli, M.; Vaccarella, R.; Chieppa, G. Biomasse benthoniche nel Mar Piccolo di Taranto e loro rapporto con i banchi naturali di mitili. *Nat. Sicil.* **1982**, *2*, 263–268.

67. Brunetti, R.; Mastrototaro, F. The non-indigenous stolidobranch ascidian *Polyandrocarpa zorritensis* in the Mediterranean: Description, larval morphology and pattern of vascular budding. *Zootaxa* **2004**, *528*, 1–8. [CrossRef]

68. Mastrototaro, F.; Brunetti, R. The non-indigenous ascidian *Distaplia bermudensis* in the Mediterranean: Comparison with the native species *Distaplia magnilarva* and *Distaplia lucillae* sp. nov. *J. Mar. Biol. Assoc. UK* **2006**, *86*, 181–185. [CrossRef]

69. Longo, C.; Mastrototaro, F.; Corriero, G. Occurrence of *Paraleucilla magna* (Porifera: Calcarea) in the Mediterranean sea. *J. Mar. Biol. Assoc. UK* **2007**, *87*, 1749–1755. [CrossRef]

70. Pierri, C.; Longo, C.; Giangrande, A. Variability of fouling communities in the Mar Piccolo of Taranto (Northern Ionian Sea, Mediterranean Sea. *J. Mar. Biol. Assoc. UK* **2010**, *90*, 159. [CrossRef]

71. Petrocelli, A.; Cecere, E.; Verlaque, M. Alien marine macrophytes in transitional water systems: New entries and reappearances in a Mediterranean coastal basin. *Bioinvasions Rec.* **2013**, *2*, 177–184. [CrossRef]

72. Giangrande, A.; Licciano, M.; Lezzi, M.; Pierri, C.; Caruso, L.P.G. Allochthonous *Branchiomma* species (Anellida, Sabellidae) in the Mediterranean Sea. A case of study in the Mar Grande of Taranto. *Biol. Mar. Medit.* **2014**, *21*, 93–96.

73. Lezzi, M.; Giangrande, A. Seasonal and bathymetric effects on macrofouling invertebrates' primary succession in a Mediterraenan non-indigenous species hotspot area. *Mediterr. Mar. Sci.* **2018**, *19*, 568–584. [CrossRef]

74. Lezzi, M.; Del Pasqua, M.; Pierri, C.; Giangrande, A. Seasonal non-indigenous species succession in a marine macrofouling invertebrate community. *Biol. Invasions* **2018**, *20*, 937–961. [CrossRef]

75. Giangrande, A.; Licciano, M.; Pagliara, P.; Gambi, M. Gametogenesis and larval development in *Sabella spallanzanii* (Polychaeta, Sabellidae) from Mediterranean Sea. *Mar. Biol.* **2000**, *136*, 847–861. [CrossRef]

76. Del Pasqua, M.; Schulze, A.; Tover-Hernàndez, M.; Keppel, E.; Lezzi, M.; Gambi, M.C.; Giangrande, A. Clarifying the taxonomic status of the alien species *Branchiomma bairdi* and *Branchiomma boholense* (Annelida: Sabellidae) using molecular and morphological evidence. *PLoS ONE* **2018**, *13*, e0197104. [CrossRef] [PubMed]

77. Giangrande, A.; Pierri, C.; Del Pasqua, M.; Gravili, C.; Gambi, M.C.; Gravina, M.F. Mediterranean in check: Biological invasions in a changing sea. *Mar. Ecol.* **2020**, *41*, e12583. [CrossRef]

78. Rovere, A.; Parravicini, V.; Firpo, M.; Morri, C.; Nike Bianche, C. Combining geomorphologic, biological and accessibility values for marine natural heritage evaluation and conservation. *Aquat. Conserv.* **2011**, *21*, 541–552. [CrossRef]

79. Bianchi, C.N.; Morri, C.; Chiantore, M.; Montefalcone, M.; Parravicini, V.; Rovere, A. Mediterranean Sea biodiversity between the legacy from the past and a future of change. In *Life in the Mediterranean Sea: A Look at Habitat Changes*; Stambler, N., Ed.; Bar Ilan University: Ramat Gan, Israel, 2012; Volume 1, pp. 1–55.

80. Gatti, G.; Bianchi, C.N.; Parravicini, V.; Rovere, A.; Peirano, A.; Montefalcone, M.; Massa, F.; Morri, C. Ecological change, sliding baselines and the importance of historical data: Lessons from combining observational and quantitative data on a temperate reef over 70 years. *PLoS ONE* **2015**, *10*, e0118581. [CrossRef]

81. Balaguer, L.; Escudero, A.; Martín-Duque, J.F.; Mola, I.; Aronson, J. The historical reference in restoration ecology: Re-defining a cornerstone concept. *Biol. Conserv.* **2014**, *176*, 12–20. [CrossRef]

82. Kahn Jr, P.H.; Friedman, B. Environmental views and values of children in an inner-city black community. *Child. Dev.* **1995**, *66*, 1403–1417. [CrossRef]

83. Pauly, D. Anecdotes and shifting baseline syndrome of fisheries. *Trends Ecol. Evol.* **1995**, *10*, 430. [CrossRef]

84. Pinnegar, J.K.; Engelhard, G.H. The 'shifting baseline' phenomenon: A global perspective. *Rev. Fish Biol. Fish.* **2008**, *18*, 1–16. [CrossRef]

85. Papworth, S.K.; Rist, J.; Coad, L.; Milner-Gulland, E.J. Evidence for shifting baseline syndrome in conservation. *Conserv. Lett.* **2009**, *2*, 93–100. [CrossRef]

86. Lo Brutto, S. A finding at the Natural History Museum of Florence affords the holotype designation of *Orchestia stephenseni* Cecchini, 1928 (Crustacea: Amphipoda: Talitridae). *Zootaxa* **2017**, *4*, 569–572. [CrossRef] [PubMed]

87. Lo Brutto, S. The case of a rudderfish highlights the role of natural history museums as sentinels of bio-invasions. *Zootaxa* **2017**, *3*, 382–386. [CrossRef] [PubMed]

88. Iaciofano, D.; Lo Brutto, S. *Parhyale plumicornis* (Crustacea: Amphipoda: Hyalidae): Is this an anti-lessepsian Mediterranean species? Morphological remarks, molecular markers and ecological notes as tools for future records. *Syst. Biodiver.* **2017**, *15*, 238–252. [CrossRef]

89. Lo Brutto, S.; Iaciofano, D. A taxonomic revision helps to clarify differences between the Atlantic invasive *Ptilohyale littoralis* and the Mediterranean endemic *Parhyale plumicornis* (Crustacea, Amphipoda). *ZooKeys* **2018**, *754*, 47–62. [CrossRef]

90. Bellia, E.; Cesara, G.; Cigna, V.; Lo Brutto, S.; Massa, B. *Epinephelus sicanus* (Doderlein, 1882) (Perciformes: Serranidae: Epinephelinae), a valid species of grouper from the Mediterranean Sea. *Zootaxa* **2020**, *4758*, 191–195. [CrossRef]

91. Giangrande, A.; Licciano, M.; Lezzi, M.; Caruso, L.; Musco, L.; Miglietta, A.M. La collezione degli Anellidi Policheti del Museo di Biologia Marina "Pietro Parenzan", Università del Salento. *Museol. Sci.* **2015**, *9*, 52–56.

92. Hoeksema, B.W.; Koh, E.G.L. Depauperation of the mushroom coral reef (Fungiidae) of Singapore (1860s–2006) in changing reef conditions. *Raffles Bull. Zool. Suppl.* **2009**, *22*, 91–101.

93. Hoeksema, B.W.; van der Land, J.; van der Meij, S.E.T.; van Ofwegen, L.P.; Reijnen, B.T.; Rob, W.M.; van Soest, R.W.M.; de Voogd, N.J. Unforeseen importance of historical collections as baselines to determine biotic change of coral reefs: The Saba Bank case. *Mar. Ecol.* **2011**, *32*, 135–141.

94. Baisre, J.A. Shifting baselines and the extinction of the Caribbean monk seal. *Conserv. Biol.* **2013**, *27*, 927–935. [CrossRef]

95. Leonetti, F.L.; Sperone, E.; Travaglini, A.; Mojetta, A.R.; Signore, M.; Psomadakis, P.N.; Dinkel, T.M.; Bottaro, M. Filling the gap and improving conservation: How IUCN Red Lists and historical scientific data can shed more light on threatened sharks in the Italian seas. *Diversity* **2020**, *12*, 289. [CrossRef]

96. Giangrande, A. Biodiversity, conservation and the 'Taxonomic impediment'. *Aquat. Conserv.* **2003**, *13*, 451–459. [CrossRef]

97. Boero, F. The study of species in the Era of Biodiversity: A tale of stupidity. *Diversity* **2010**, *2*, 115–126. [CrossRef]

98. Boero, F. Light after dark: The partnership for enhancing expertise in taxonomy. *Trends Ecol. Evol.* **2001**, *16*, 266. [CrossRef]

99. Fanelli, G.; Portacci, G.; Boero, F. La variabilità del benthos di Porto Cesareo (LE) (Mar Ionio) attraverso l'analisi delle serie storiche (1989–2004). *Biol. Mar. Medit.* **2006**, *13*, 71–77.

Publisher's Note: MDPI stays neutral with regard to jurisdictional claims in published maps and institutional affiliations.

 *diversity*

Article

Diversity and Distribution Patterns of Hard Bottom Polychaete Assemblages in the North Adriatic Sea (Mediterranean)

Barbara Mikac [1,*], Margherita Licciano [2], Andrej Jaklin [3], Ljiljana Iveša [3], Adriana Giangrande [2] and Luigi Musco [4]

[1]   Interdepartmental Research Centre for Environmental Sciences, University of Bologna, Via Sant'Alberto 163, 48123 Ravenna, Italy

[2]   Department of Biological and Environmental Sciences and Technologies, University of Salento, Via Provinciale Lecce-Monteroni, 73100 Lecce, Italy; margherita.licciano@unisalento.it (M.L.); adriana.giangrande@unisalento.it (A.G.)

[3]   Center for Marine Research, Ruđer Bošković Institute, Giordano Paliaga 5, 52210 Rovinj, Croatia; jaklin@cim.irb.hr (A.J.); ivesa@cim.irb.hr (Lj.I.)

[4]   Stazione Zoologica Anton Dohrn, Integrative Marine Ecology Department, Villa Comunale, 80121 Naples, Italy; luigi.musco@szn.it

*   Correspondence: barbara.mikac@unibo.it

Received: 3 September 2020; Accepted: 19 October 2020; Published: 21 October 2020

Abstract: The knowledge on the hard bottom polychaete assemblages in the Northern Adriatic Sea, a Mediterranean region strongly affected by environmental pressures, is scarce and outdated. The objective of this paper was to update the information on polychaete diversity and depict their patterns of natural spatial variation, in relation to changes in algal coverage at increasing depth. Hard bottom benthos was quantitatively sampled by scraping off the substrate from three stations at Sveti Ivan Island (North Adriatic) at three depths (1.5 m, 5 m and 25 m). Polychaete fauna comprised 107 taxa (the majority of them identified at species level) belonging to 22 families, with the family Syllidae ranking first in terms of number of species, followed by Sabellidae, Nereididae, Eunicidae and Serpulidae. Considering the number of polychaete species and their identity, the present data differed considerably from previous studies carried out in the area. Two alien species, *Lepidonotus tenuisetosus*, which represented a new record for the Adriatic Sea, and *Nereis persica*, were recorded. The highest mean abundance, species diversity and internal structural similarity of polychaete assemblages were found at 5 m depth, characterised by complex and heterogeneous algal habitat. The DISTLM forward analysis revealed that the distribution of several algal taxa as well as some algal functional-morphological groups significantly explained the observed distribution patterns of abundance and diversity of polychaete assemblages. The diversity of the North Adriatic hard bottom polychaete fauna is largely underestimated and needs regular updating in order to detect and monitor changes of benthic communities in the area

Keywords: Annelida; Polychaeta; benthos; community structure; algae

1. Introduction

The North Adriatic Sea is the northernmost sector of the Mediterranean Sea, with peculiar geomorphological, hydrographical and biogeographical characteristics. With an average depth of 35 m, this semienclosed basin represents the most extensive region of shallow water in the Mediterranean [1,2], being one of the most productive areas too, due to high amount of nutrients loaded by the Po River [3]. It is also the coldest Mediterranean sector, together with the Gulf of Lion and the North Aegean

Sea, thus inhabited by species of boreal affinity [4,5]. The Adriatic Sea and in particular its northern part, exhibits the highest species richness of invertebrates in the Mediterranean basin [6]. Being also densely populated and thus under high anthropogenic pressures, the North Adriatic is a sensitive area currently undergoing severe environmental changes (climate change, fishing impacts, destruction of habitats, pollution, introduction of non indigenous species) that affect the benthic communities [6–8]. Updating the knowledge about benthic diversity and understanding patterns of benthic assemblages' vertical and horizontal spatial variation are benchmarks for detecting and monitoring environmental changes, also according to the European Union Marine Strategy Framework Directive [9].

Polychaetes are among the most abundant and species-rich marine benthic groups, showing a wide functional diversity and adaptation to different environmental conditions [10,11]. Thus, they are often used as surrogates to estimate the state and dynamics of benthic communities [12–15]. North Adriatic soft bottom polychaete fauna is well known, e.g., [16–32], while the knowledge on hard bottom polychaete assemblages from natural substrates is scarce, i.e., [33–42] and their diversity might be largely underestimated as indicated by recent studies dealing with Syllidae and Sabellidae polychaete families [43,44].

The structure of hard bottom benthic assemblages is characterised by having high small- and middle-scale spatial variability (i.e., patchiness), both alongshore and at different depths, which is caused by various interplaying biological (e.g., predation, competition, recruitment) and physical-chemical (e.g., light intensity, temperature, salinity, hydrodynamics, sedimentation, habitat complexity) factors [45–50]. In particular, the distribution of hard-bottom polychaetes is strongly dependent on the bathymetric variation in algal composition and the associated changes in algal forms [51–53]. However, studies aiming at understanding the role of the above-mentioned factors in structuring polychaete assemblages have not been done in the North Adriatic so far.

The aims of our study were: (1) to update the knowledge of the faunal composition of the North Adriatic hard bottom polychaete assemblages, and (2) to assess variation of their spatial distribution in relation to changes in algal assemblages along a bathymetric gradient and according to substrate orientation.

2. Materials and Methods

The study area was in the vicinity of the city of Rovinj (Croatia, North Adriatic Sea) at Sveti Ivan Island. Islands of the Rovinj archipelago and the coastal area of up to 500 m from the coastline, were proclaimed by Rovinj Municipality a natural landscape reserve. The area is characterized by calcareous rocky shelf extending from 0 to about 25 m depth, with a gentle-medium slope. The submarine slopes of the Sveti Ivan Island are representing typical infralittoral environments of the North Adriatic Shelf.

Collecting surveys were carried out in June 2007, taking into consideration that the maximum development of macroalgal assemblages in the Northern Adriatic Sea occurs in the spring–early summer period [54,55]. Benthos was sampled using scuba diving at stations A (N 45° 02.69′, E 13° 37.18′) and B (N 45° 02.7′, E 13° 37.48′) on the southern side and station C (N 45° 02.87′, E 13° 37.34′) on the northern side of the Island (Figure 1). At each station, three depths (1.5 m, 5 m and 25 m) were appointed along a vertical transect and at each depth three replicates of 10 × 10 cm surface quadrats covered with macroalgae were randomly chosen (27 samples in total). Samples were collected by scraping off the substrate including the whole algal coverage present within the 10 × 10 cm quadrats using hammer and chisel. The scraped material was collected within plastic bags. Although ordinarily a sampling area of 20 × 20 cm is suggested for benthic studies in the Mediterranean [56], the small sample size (10 × 10 cm) in this research was chosen in order to minimize sampling impacts in the natural reserve. Furthermore, 10 × 10 cm replicate areas were already used in other studies dealing with polychaetes in the Mediterranean Sea and revealed to allow acceptable representation of polychaete diversity and distribution patterns [11,57,58]. Each replicate unit (10 × 10 cm surface quadrats) was photographed underwater to facilitate the determination of the associated algal taxa and their percent coverage.

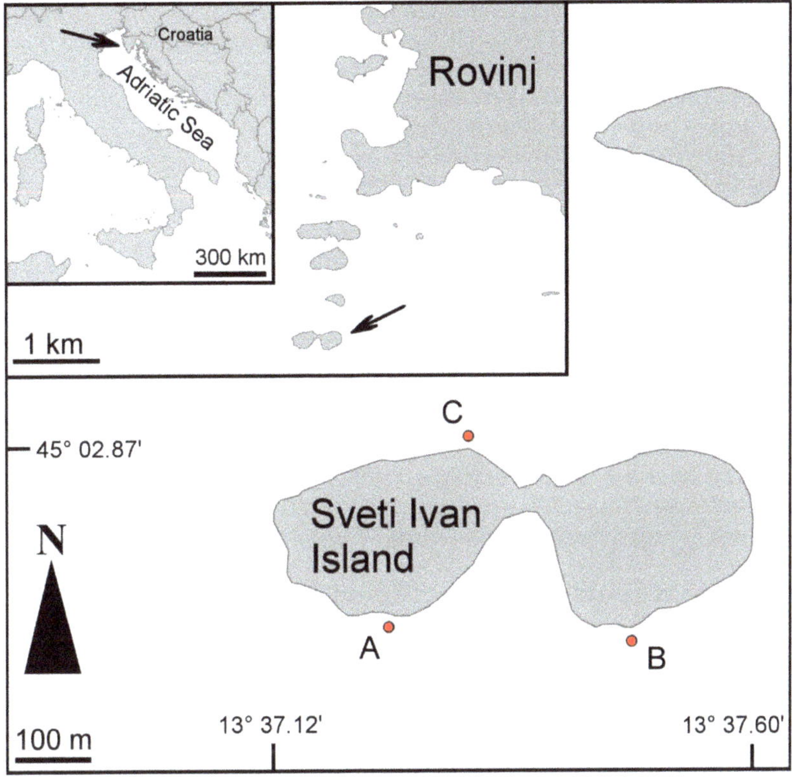

Figure 1. Location of the study area and sampling stations (**A–C**). Arrows indicating city of Rovinj area and Sveti Ivan Island.

In the laboratory, samples were fixed in 8% formaldehyde seawater solution and subsequently rinsed with fresh water and sieved through 0.5mm mesh and preserved in 70% ethanol. Polychaetes and algae were sorted and determined to the lowest taxonomic level possible using stereo- and light microscope. In some cases, it was not possible to identify the organisms at species level but only at higher taxonomic categories (e.g., genus, family). For this reason, we used the term taxa instead of species throughout the manuscript to indicate the recorded taxonomic entities. When specimens belonging to the same genus were clearly different from each other, however, we considered them belonging to different undetermined species (e.g., *Nereis* sp. 1 and *Nereis* sp. 2). The coverage of each algal thallus, representing the surface covered in an orthogonal projection, was determined according to Boudouresque [56] and Cormaci et al. [59]. Algae from each sample were placed on the surface that has an area equal to that sampled in situ (10 × 10 cm). Consequently, for each algal species the percentage of total quadrat area (100 cm^2) covered by the projection of all the thalli was estimated. Value for total algal coverage can reach more than 100% in the presence of multilayered assemblages (as canopy forming algae) or epibiosis [56]. Additionally, algae were grouped in the following functional-morphological groups: Articulated calcareous, Corticated, Encrusting, Filamentous, Foliose and Leathery [60,61], and coverage of each functional-morfological group per sample was calculated as well. The polychaete specimens were deposited at the Center for Marine Research of the Ruđer Bošković Institute in Rovinj (Croatia) (IRB-CIM) and in the collection of the Natural History Museum in Rijeka (Croatia) (PMR).

For each replicate sample, polychaete assemblages were characterised by their respective abundance (*N*), species richness (*S*), Hill's species diversity index (*N1*) and Hill's evenness index (*N10*) [62]. In order to graphically represent trends in the number of species found within the collected samples, species accumulation curve of observed species (Sobs) was created. Moreover, to estimate the number of species potentially present in the area, curves of estimated number of species were calculated using the Jacknife 1, Jacknife 2 and Bootstrap methods [63].

Nonparametric distance-based permutational analysis of variance by permutation of residuals under a reduced model (PERMANOVA) [64,65] was used to test for differences in univariate indices (based on Euclidean distances of untransformed data) and in multivariate structure (based on Bray–Curtis similarity of untransformed data) of the polychaete assemblages between stations and depths. PERMANOVA design included two crossed factors: station (3 levels, random) and depth (3 levels, fixed). Post-hoc pair-wise comparisons allowed detecting the source of significant variations. For significant terms, a permutational analysis of multivariate dispersions (PERMDISP) [66] was used to test the homogeneity of samples dispersion from their group centroids. When the number of permutations was low (less than 1000), Monte Carlo probability (P(MC)) was considered instead of permutational probability (P(perm)). To calculate *p* values for PERMANOVA and PERMDISP, 9999 permutations were used.

Multivariate patterns were visualised by nonmetric multidimensional scaling ordination (nMDS). The similarity percentage routine (SIMPER) [67] (70% cut off), was used to detect the taxa most responsible for within-depth similarity and between-depth dissimilarity and, at each depth, the between-station dissimilarity. When analysing dissimilarity between stations, taxa were considered important if they exceeded an arbitrarily chosen threshold of 4% of dissimilarity between stations at each depth.

Potential relationships among structuring algal taxa, algal functional-morphological groups, depth and orientation of the sampling station in respect to the island geography (south/north)) and the variation of polychaete assemblages were assessed by nonparametric multiple regression analyses based on Bray–Curtis dissimilarities [65] exerting the distance-based multivariate analysis for a linear model using forward selection procedure (DISTLM), with 9999 permutations. Resemblance matrix produced by DISTLM analyses informed on the correlation among all pairs of explanatory variables to check for multicolinearity [68]. The predictor variables included depth, orientation and percent coverage of each structuring algal species (cut-off 5% of the total cover) in the first DISTLM analysis and depth, orientation and percent coverage of algal functional-morphological groups in the second one. Results of the forward selection procedure with the sequential tests (i.e., fitting each variable one at a time, conditional on the variables that are already included in the model) were presented. All analyses were done using PRIMER v.6 [69], with the add-on PERMANOVA+ [66].

3. Results

3.1. Algal Assemblages

Forty-eight algal taxa were recorded in the research area (Table S1). Total mean algal coverage was higher at 1.5 m and 5 m depths if compared to 25 m depth in all three stations (Figure 2). At 1.5 m depth, algal assemblages were characterised by the high coverage of the articulated calcareous algae (mostly *Corallina officinalis* Linnaeus, *Haliptilon* sp. and *Jania* spp.), the corticated algae (mostly *Alsidium* sp. and *Laurencia obtusa* (Hudson) J.V.Lamouroux) and the foliose algae (mostly *Padina pavonica* (Linnaeus) Thivy, *Dictyota dichotoma* (Hudson) J.V.Lamouroux and *Dictyota dichotoma* var. *intricata* (C.Agardh) Greville) (Figure 3a). Filamentous algae (mostly *Ectocarpus* sp. and *Carradoriella elongata* (Hudson) A.M.Savoie & G.W.Saunders) and encrusting algae (mostly *Valonia utricularis* (Roth) C.Agardh and *Peyssonnelia rubra* (Greville) J.Agardh) were present with low percent coverage at 1.5 m depth. At 5 m depth, encrusting algae (mostly *P. rubra* and *P. heteromorpha* (Zanardini) Athanasiadis) dominated, followed by foliose algae (mostly *P. pavonica*, *Flabellia petiolata* (Turra) Nizamuddin, 1987 and *D. dichotoma*), while articulated calcareous algae (mostly *C. officinalis*, *Haliption* sp. and *Jania* spp) were present with a lower coverage (Figure 3b). Corticated algae (mostly *Gelidium spinosum* (S.G.Gmelin) P.C.Silva) and filamentous algae (mostly *Polysiphonia* sp. and *Cladophora* spp.) had very low percent coverage at 5 m depth. At 25 m depth, filamentous algae (mostly *Cladophora* spp., *Sphacelaria plumula* Zanardini, and *Polysiphonia* sp.) were prevalent and followed by encrusting algae (mostly *P. rubra* and an unidentified rose coloured encrusting algae) (Figure 3c). Foliose algae (mostly *Rhodymenia* sp., *D. dichotoma* and unidentified foliose algae) showed a very low coverage at 25 m depth. At this depth, corticated algae (mostly *Rodriguezella* sp.) were present only at stations B and C, while articulated calcareous algae (only *Halimeda tuna* (J.Ellis & Solander) J.V.Lamouroux) were found only at station C, all of them with very low coverage. *Cystoseira compressa* (Esper) Gerloff & Nizamuddin, the single species of the leathery functional group, appeared with a 2% coverage only in one sample from 1.5 m depth (Figure 2).

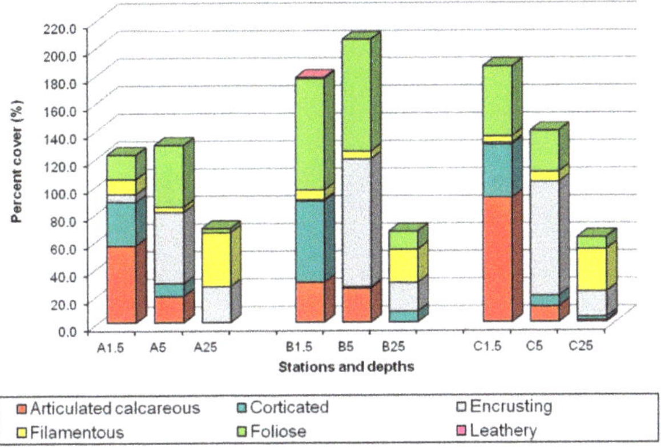

Figure 2. Mean total percent coverage of algae and algal functional-morphological groups, per stations and depths.

Figure 3. Representative algal assemblages at: (**a**) 1.5 m depth; (**b**) 5 m depth; (**c**) 25 m depth.

3.2. Composition and Diversity of Polychaete Assemblages

As a whole, 1993 polychaete specimens from 107 taxa (94 at species level) belonging to 22 families were found (Table S2). The richest families in terms of number of species were Syllidae (39 species), Sabellidae (13), Nereididae (12), Eunicidae (7) and Serpulidae (7), while eleven families were represented by only one species. Altogether 62 species from 16 families were found at 1.5 m depth, 73 species from 15 families at 5 m depth and 66 species from 15 families at 25 m depth.

The mean number of individuals and species, as well as the mean Hill's species diversity index, were the highest at 5 m depth, while no particular pattern could be observed in the Hill's evenness index (Figure 4a,b). Significant differences in species richness and Hill's species diversity index were revealed both horizontally (between stations) and vertically (between depths), while differences in polychaete abundance and Hill's evenness index were significant only between stations (Table 1). In particular, the assemblages at 5 m depth had significantly higher species richness and Hill's species diversity index than those of 1.5 m and 25 m depths, at most of the stations (Table S3). Significant differences between stations in all univariate diversity descriptors were revealed mostly at 1.5m depth and to a lesser extent also at 25 m depth (Table S3).

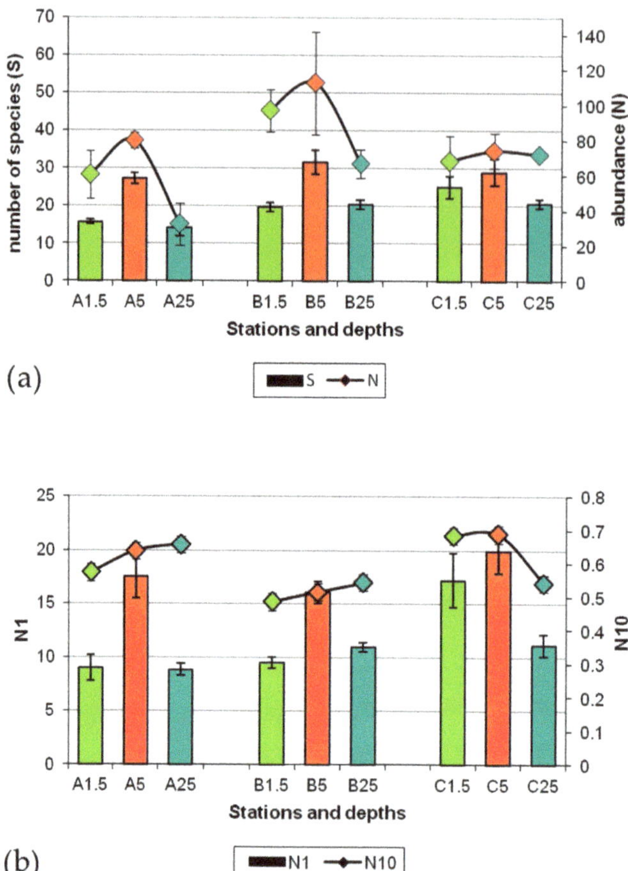

Figure 4. (**a**) Mean (±SE) total abundance (N; lines) and species richness (S; bars); (**b**) Mean (±SE) Hill's species diversity index (N1; bars) and Hill's evenness index (N10; lines) at each studies station and depth.

Syllidae was the dominant family at all three depths, both in abundance and species richness, followed by Nereididae, Sabellidae and Eunicidae at 1.5 m depth and, at 5m depth, by Sabellidae, Nereididae and Eunicidae considering abundance, and Nereididae, Eunicidae and Sabellidae considering species richness (Figure 5). At 25 m depth, Syllidae were particularly dominant, and followed by Sabellidae, Nereididae, Serpulidae and Eunicidae, both considering abundance and number of species. Abundance and diversity of Serpulidae increased with depth. In fact, at 1.5 m only one specimen of *Vermiliopsis infundibulum* (Philippi, 1844) was found, while the Serpulidae were represented by 18 specimens belonging to 4 species at 5 m depth and 23 specimens belonging to 6 species at 25 m depth.

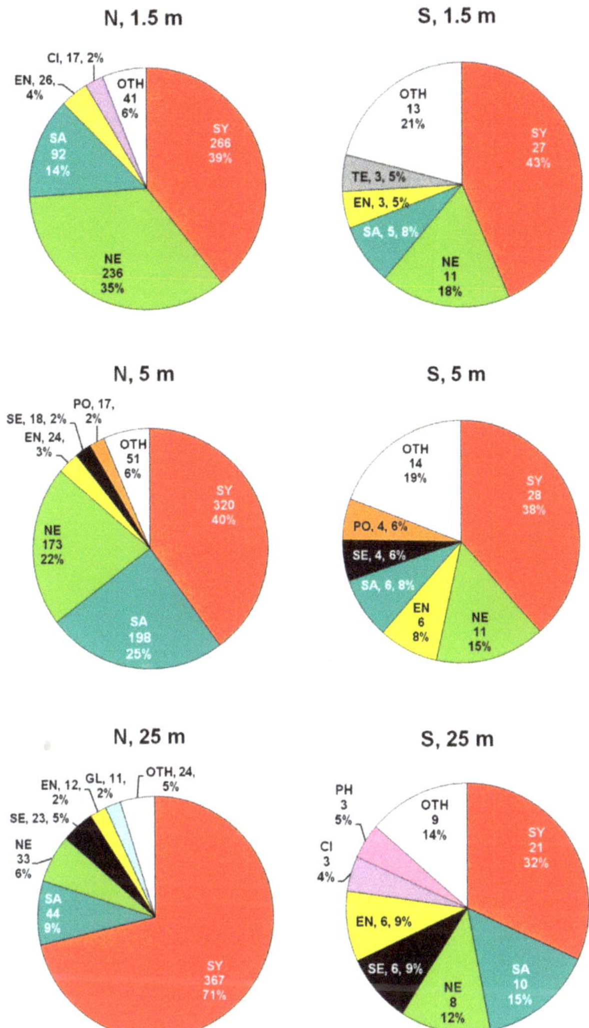

Figure 5. Cumulative abundance (N; on the left side) and species richness (S; on the right side) of polychaetes per family at the three studied depths. CI—Cirratulidae, EN—Eunicidae, GL—Glyceridae, NE—Nereididae, PH—Phyllodocidae, PO—Polynoidae, SA—Sabellidae, SE—Serpulidae, SY—Syllidae, TE—Terebellidae, OTH—other families.

Most species were locally rare, with 36 species found only in one sample (Figure S1), while only *Sphaerosyllis pirifera* Claparède, 1868, *Amphiglena mediterranea* (Leydig, 1851) and *Platynereis dumerilii* (Audouin & Milne Edwards, 1833) were widely distributed in the whole area in most of the samples (25, 21 and 21 respectively). Neither the accumulation, nor the estimator curves (Jacknife1, Jacknife2, Bootstrap) reached the asymptote (Figure S2), suggesting a potential higher number of species ranging from 121 to 167 in the area.

Table 1. Results of PERMANOVA and PERMDISP analyses (untransformed data) testing for differences in abundance (N), species richness (S), Hill's species diversity index (N1), Hill's evenness index (N10) and structure (Stru) of polychaete assemblages between stations (st) and depths (de). df, degrees of freedom; SS, sum of squares; MS, mean squares; Up, unique perms; F, F-ratio; P (perm), probability. Significant p-values ($p < 0.05$) are given in bold.

	Source	PERMANOVA						PERMDISP			
		df	SS	MS	Pseudo-F	Up	P (perm)	df1	df2	F	P (perm)
N	st	2	5442.3	2721.1	4.7539	9953	**0.0192**	2	24	2.4351	0.1038
	de	2	4607.2	2303.6	4.3316	6066	0.0995			-	-
	stxde	4	2127.3	531.81	0.92908	9956	0.4792			-	-
	Res	18	10303	572.41							
S	st	2	178.74	89.37	6.4866	9949	**0.0076**	2	24	0.3413	0.8379
	de	2	627.63	313.81	18.976	3825	**0.0254**	2	24	0.2366	0.7997
	stxde	4	66.148	16.537	1.2003	9950	0.3431			-	-
	Res	18	248	13.778							
N1	st	2	101.95	50.974	7.9594	9955	**0.0028**	2	24	1.0783	0.4925
	de	2	282.77	141.38	9.7949	6091	**0.0459**	2	24	3.2631	0.0885
	stxde	4	57.737	14.434	2.2538	9950	0.0995			-	-
	Res	18	115.28	6.4043							
N10	st	2	0.0809	0.0405	4.293	9950	**0.0304**	2	24	1.1321	0.4103
	de	2	0.0065	0.0032	0.25075	6086	0.7786			-	-
	stxde	4	0.0517	0.0129	1.3718	9937	0.2792			-	-
	Res	18	0.1696	0.0094							
Stru	st	2	7282.2	3641.1	2.5077	9900	**0.0001**	2	24	4.2513	0.0549
	de	2	26808	13404	4.3784	6114	**0.0176**	2	24	2.4735	0.1668
	stxde	4	12246	3061.4	2.1084	9843	**0.0001**	8	18	8.0471	**0.0122**
	Res	18	26136	1452							

3.3. Patterns of Variation of Polychaete Assemblages Structure

There were significant differences in the structure of polychaete assemblages both among stations and depths, but also for the interaction term station × depth (Table 1), with the significant alongshore variation occurring only at 1.5 m depth, between stations B and A and between stations B and C (Table S3). Moreover, significant differences in the structure of polychaete assemblages were revealed at all stations between 1.5 m and 25 m depth, and between 5 m and 25 m depth, while differences between 1.5 m and 5 m depth were significant only at station B. Results of PERMDISP analyses confirmed that these differences were not barely due to differences in the dispersion of the samples (Table S3).

At all stations, difference in polychaete assemblages between three depths was clearly evident, with assemblages from 1.5 m and 5 m depths being more similar among each other and different from those at 25 m depth (Figure 6). Station A showed the highest scattered distribution among replicates at 25m depth, and the most homogeneous assemblages were those at 5m depth. In fact, the average similarity in species composition and abundance between samples was the highest at 5 m depth (45.22%), intermediate at 25 m depth (39.51%) and the lowest at 1.5 m depth (34.39%) (Table 2A). The most abundant species were the sabellid *Amphiglena mediterranea*, the nereidids *Platynereis dumerilii*, *Nereis usticensis* Cantone, Catalano & Badalamenti, 2003 and *Nereis pulsatoria* (Savigny, 1822) and the syllids *Syllis rosea* (Langerhans, 1879), *S. pirifera* and *Exogone dispar* (Webster, 1879) at 1.5m depth; the sabellid *A. mediterranea*, the syllids *S. pirifera*, *E. dispar*, *Syllis variegata* Grube, 1860, *Syllis prolifera* Krohn, 1852 and *Syllis corallicola* Verrill, 1900, and nereidids *P. dumerilii*, *Nereis* sp. 1 and unidentified juvenile nereidids at 5m depth; and the syllids *Syllis armillaris* (O.F. Müller, 1776), *Syllis gracilis* Grube, 1840, *Syllis gerlachi* (Hartmann-Schröder, 1960) and *S. pirifera* at 25 m depth (Table 2A). The lowest average dissimilarity in species composition and abundance was between 1.5 m and 5 m depth (68.67%), while it was higher between 1.5 m and 25 m (88.42%) and between 5 m and 25 m depth (79.59%) (Table 2B). Differences in abundance of the most abundant species of Syllidae (i.e., *S. rosea*, *S. prolifera*, *S. armillaris*,

S. gracilis, S. gerlachi, S. pirifera), Nereididae (i.e., *N. usticensis, P. dumerilii, Nereis* sp. 1, Nereididae juv. indet.) and Sabellidae (i.e., *A. mediterranea*) at different depths, were mainly responsible for these dissimilarities (Table 2B). *Amphiglena mediterranea* had the highest abundance at 5 m depth, was slightly less abundant at 1.5 m depth and poorly represented at 25 m depth. *Syllis armillaris* and *S. gracilis* were very abundant at 25 m depth and poorly represented at 1.5 and 5 m depth. Only low percentages of dissimilarities were due to the differences in taxonomic composition (i.e., presence/absence of species). Namely, the sabellid *Amphicorina rovignensis* Mikac, Giangrande & Licciano, 2013 characterised 1.5 m depth but was absent at 5 m depth, *N. usticensis* and *S. rosea* characterised 1.5 m depth but were absent at 25 m depth, the syllid *Odontosyllis ctenostoma* Claparède, 1868 characterised 5 m depth, but was absent at 25 m depth, and the sabellid *Hypsicomus stichophthalmos* (Grube, 1863) characterised 25 m depth but was absent at 5 m depth (Table 2B).

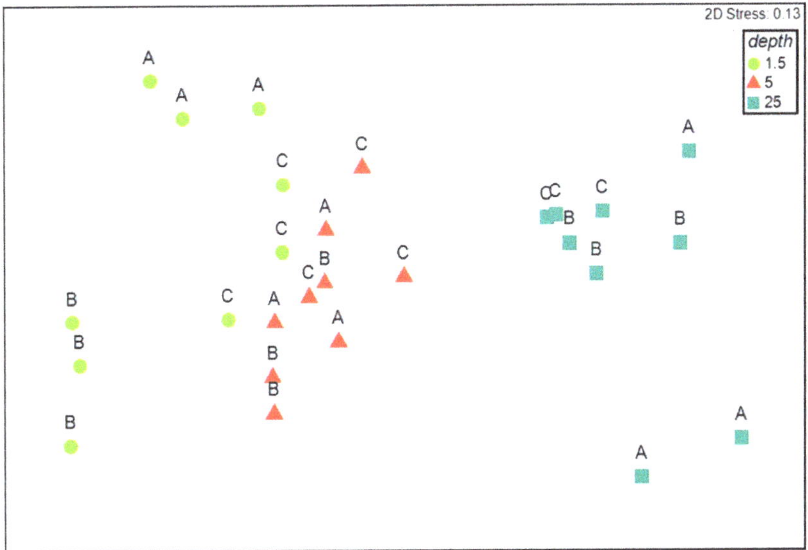

Figure 6. Nonmetric multidimensional scaling (MDS) ordination plot based on Bray–Curtis similarity of untransformed data, comparing structure of polychaete assemblages between samples. **A–C** = stations.

The species that contributed with 4% or more to the dissimilarity between stations at each depth, were mostly the same ones that characterised that depth (Table S4): *N. usticensis, N. pulsatoria, S. rosea, S. prolifera, A. mediterranea, S. pirifera, P. dumerilii, E. dispar, Ceratonereis (Composetia) costae* (Grube, 1840) and Nereididae juv. indet. at 1.5 m depth; *A. mediterranea, A. rovignensis, S. prolifera, S. gerlachi, Nereis* sp. 1, *S. pirifera* and *P. dumerilii* at 5 m depth; and *S. armillaris, S. pirifera, H. stichophthalmos, S. gracilis, S. variegata, S. gerlachi* and *Spirobranchus triqueter* (Linnaeus, 1758) at 25 m depth.

Table 2. Results of SIMPER analyses (cut-off 70%) used to identify taxa that mostly contribute to (**A**) faunal similarity within depths, (**B**) faunal dissimilarity between depths. Abund = mean abundance, Sim% = mean similarity, Sim/SD = similarity/standard deviation, Contrib% = contribution relative to single taxon, Cum% = cumulative contribution, Av.Ab = mean abundance, Diss% = mean dissimilarity, Diss/SD = dissimilarity/standard deviation.

	Group Sim%	Species	Abund	Sim%	Sim/ SD	Contrib%	Cum%
A							
	1.5 m 34.39%	*Amphiglena mediterranea*	9.67	8.14	1.63	23.68	23.68
		Platynereis dumerilii	6.11	5.41	1.48	15.72	39.40
		Nereis usticensis	11.00	2.45	0.33	7.13	46.54
		Syllis rosea	7.00	2.45	1.11	7.12	53.66
		Sphaerosyllis pirifera	4.67	2.43	0.71	7.06	60.72
		Nereis pulsatoria	3.11	2.35	1.05	6.85	67.56
		Exogone dispar	2.89	1.62	0.65	4.70	72.26
	5 m 45.22%	*Amphiglena mediterranea*	17.56	10.46	1.43	23.14	23.14
		Sphaerosyllis pirifera	5.67	4.06	1.46	8.97	32.11
		Nereididae juv. indet.	4.00	3.04	2.11	6.71	38.82
		Nereis sp. 1	4.56	2.92	1.37	6.46	45.28
		Exogone dispar	3.67	2.79	2.13	6.17	51.45
		Syllis variegata	3.33	2.55	1.33	5.63	57.08
		Platynereis dumerilii	3.56	2.50	1.17	5.52	62.60
		Syllis prolifera	5.33	2.12	0.71	4.69	67.29
		Syllis corallicola	3.11	2.08	2.29	4.59	71.88
	25 m 39.51%	*Syllis armillaris*	13.22	13.05	1.55	33.04	33.04
		Syllis gracilis	6.67	8.38	1.88	21.21	54.25
		Sphaerosyllis pirifera	8.78	5.93	0.98	15.01	69.27
		Syllis gerlachi	4.11	3.65	1.44	9.24	78.50

	Groups Diss%	Species	Av. Ab Group 1.5	Av.Ab Group 5	Diss%	Diss/SD	Contrib%	Cum%
B								
	1.5 and 5 68.67%	*Amphiglena mediterranea*	9.67	17.56	7.34	1.35	10.69	10.69
		Nereis usticensis	11.00	0.11	5.99	0.71	8.72	19.41
		Syllis rosea	7.00	0.44	4.01	0.58	5.83	25.24
		Syllis prolifera	5.44	5.33	3.89	1.16	5.66	30.91
		Sphaerosyllis pirifera	4.67	5.67	3.38	1.13	4.92	35.83
		Platynereis dumerilii	6.11	3.56	2.53	1.21	3.68	39.51
		Nereis sp. 1	1.11	4.56	2.39	1.22	3.48	42.99
		Nereididae juv. indet.	2.00	4.00	2.16	1.47	3.15	46.14
		Syllis variegata	0.33	3.33	2.01	1.44	2.92	49.07
		Exogone dispar	2.89	3.67	1.95	1.29	2.84	51.90
		Syllis corallicola	1.00	3.11	1.89	1.69	2.75	54.65
		Syllis gerlachi	0.22	3.00	1.88	0.76	2.74	57.39
		Nereis pulsatoria	3.11	1.00	1.62	1.18	2.35	59.74
		Amphicorina rovignensis	0.00	3.33	1.46	0.41	2.13	61.87
		Nereis rava	0.78	2.67	1.46	1.30	2.13	64.00
		Ceratonereis (Composetia) costae	1.11	2.00	1.42	0.92	2.06	66.07
		Odontosyllis ctenostoma	0.33	2.44	1.36	0.86	1.98	68.05
		Lysidice unicornis	2.33	0.44	1.14	1.11	1.66	69.71
		Syllis armillaris	1.00	1.89	1.11	0.98	1.62	71.33
			Group 1.5	Group 25				
	1.5 and 25 88.42%	*Syllis armillaris*	1.00	13.22	9.06	1.75	10.24	10.24
		Amphiglena mediterranea	9.67	0.33	7.29	1.53	8.24	18.48
		Nereis usticensis	11.00	0.00	7.20	0.70	8.14	26.63
		Sphaerosyllis pirifera	4.67	8.78	6.06	1.13	6.85	33.48
		Syllis rosea	7.00	0.00	5.26	0.61	5.95	39.43
		Syllis gracilis	0.22	6.67	5.09	1.73	5.75	45.18
		Platynereis dumerilii	6.11	0.44	4.62	1.26	5.23	50.41
		Syllis prolifera	5.44	1.00	3.82	0.86	4.32	54.73
		Syllis gerlachi	0.22	4.11	2.87	1.31	3.24	57.97
		Hypsicomus stichophthalmos	0.11	3.11	2.47	0.43	2.79	60.76
		Nereis pulsatoria	3.11	0.11	2.45	1.26	2.77	63.54
		Exogone dispar	2.89	0.11	2.34	0.99	2.65	66.18
		Lysidice unicornis	2.33	0.11	1.62	1.25	1.83	68.01
		Nereididae juv. indet.	2.00	0.11	1.43	0.74	1.62	69.63
		Dodecaceria concharum	1.67	0.22	1.30	0.98	1.47	71.10
			Group 5	Group 25				

Table 2. *Cont.*

	Group Sim%	Species		Abund	Sim%	Sim/SD	Contrib%	Cum%
B								
	5 and 25	*Amphiglena mediterranea*	17.56	0.33	11.40	1.52	14.33	14.33
	79.59%	*Syllis armillaris*	1.89	13.22	7.94	1.62	9.97	24.30
		Sphaerosyllis pirifera	5.67	8.78	5.07	1.23	6.37	30.66
		Syllis gracilis	0.67	6.67	4.29	1.65	5.40	36.06
		Syllis prolifera	5.33	1.00	3.24	1.13	4.08	40.14
		Nereis sp. 1	4.56	0.56	2.89	1.26	3.63	43.77
		Nereididae juv. indet.	4.00	0.11	2.73	1.57	3.43	47.20
		Syllis gerlachi	3.00	4.11	2.64	1.11	3.32	50.52
		Exogone dispar	3.67	0.11	2.59	1.38	3.25	53.77
		Platynereis dumerilii	3.56	0.44	2.46	1.12	3.09	56.86
		Hypsicomus stichophthalmos	0.00	3.11	2.17	0.41	2.72	59.59
		Syllis variegata	3.33	1.56	1.89	1.17	2.38	61.96
		Syllis corallicola	3.11	0.44	1.75	1.43	2.20	64.17
		Amphicorina rovignensis	3.33	0.22	1.67	0.44	2.10	66.26
		Odontosyllis ctenostoma	2.44	0.00	1.56	0.85	1.96	68.22
		Nereis rava	2.67	1.11	1.55	1.16	1.95	70.17

Table 3. Results of DISTLM-forward analysis. (**A**) Variables: percent coverage of each algal taxa (5% cut-off), depth and orientation), (**B**) Variables: percent coverage of algal functional-morphological groups, depth and orientation). Only variables significantly contributing to explain variation of polychaete assemblages ($p < 0.05$) are presented. Prop.: proportion of explained variation; Cumul.: cumulative proportion of explained variation.

	Variable	SS(trace)	Pseudo-F	P	Prop.	Cumul.
	Depth	20617	9.9394	0.0001	0.28448	0.28448
A	*Peyssonnelia rubra*	5681.7	2.9532	0.0005	0.078398	0.36287
	Haliptilon sp.	5214.1	2.9279	0.0008	0.071947	0.43482
	Filamentous sp. 1	4161.5	2.488	0.001	0.057423	0.49224
	Padina pavonica	2822.7	1.7447	0.0263	0.038949	0.53119
	Codium efusum	2954.4	1.9048	0.0208	0.040766	0.57196
	Depth	20617	9.9394	0.0001	0.28448	0.28448
B	Encrusting	5292.2	2.7277	0.0007	0.073024	0.3575
	Foliose	4853.6	2.6764	0.0011	0.066971	0.42447
	Articulated calcareous	3064.3	1.7444	0.0306	0.042282	0.46675

When considering the percent coverage of each algal taxon separately, 6 variables (over a total of 36) significantly explained 57.2% of the variation of the polychaete assemblages (Table 3A). Depth explained 28.4% of the variation, while the contribution of orientation was not significant. Overall, the percent coverage of algal taxa significantly explained 28.7% of the observed variation. *Peyssonnelia rubra* (7.8%), *Haliption* sp. (7.2%), one taxon of filamentous algae (5.74%), *Padina pavonica* (3.9%) and *Codium efusum* (4.1%) were significantly related to distribution of polychaete assemblages. When considering percent coverage of algal functional-morphological groups, 4 out of 9 variables significantly explained 46.7% of the variation of the polychaete assemblages (Table 3B). Among them, depth explained 28.4%, while algal functional-morphological groups all together explained 18.2% of the variation of polychaete assemblages. Orientation was again not significant. In particular, the encrusting (7.3%), foliose (6.7%) and articulated calcareous (4.2%) algae were significantly related to the distribution of polychaete assemblages.

4. Discussion

This is the first study of the hard bottom polychaete assemblages in the North Adriatic Sea over a wide bathymetric range also applying an experimental design that allows describing patterns of spatial distribution in relation to algal coverage. Previous hard bottom studies in the area considered mostly shallower benthic habitats, using qualitative sampling methods making results hardly comparable to ours. In fact, Zavodnik [40,41] reported 38 polychaete species from *Cystoseira barbata* (Stackhouse)

C.Agardh assemblages sampled at 0.5–1 m depth and 69 species examining different brown algae from 0–3 m depths. Later, Amoureux and Katzmann [34] reported 52 species from rocky bottom algal assemblages at 6–9 m depth, while Katzmann [37] collected 93 species from *Cystoseira* assemblages sampled at different depths ranging from 0–2.5 m. Amoureux [33] found 55 polychaete species among *Cystoseira* and *Peyssonelia* algae sampled at 3 m depth. In general, we can notice higher species richness (107 taxa from 22 families) in our study, which can be explained especially by the wider depth range considered herein. Indeed, if we consider only shallower habitats, species richness found herein (62 species at 1.5 m depth and 73 species at 5 m depth) is more similar to what was previously reported for the area. Moreover, our sampling covers the most representative hard-bottom vegetated habitats in the area, supposedly allowing collecting a larger number of polychaete taxa. The macroalgal assemblages found at Sveti Ivan Island are characteristic for the western coast of the North Adriatic Sea, with photophilic algae typically dominating in shallow, and sciaphilic in deeper waters [55,70], and they can be considered as representative habitats to study the diversity of rocky bottom polychaetes in the region. The ecological status of Sv. Ivan island was previously determined using macroalgae grouped in two ecological status groups (ESG I and ESG II) according to Orfanidis et al. [71], and categorized as overall high, while moderate at 1 m depth and good at 3 m depth [54,55]. The algal assemblages herein observed corresponded to those found in the above-mentioned studies. The articulated calcareous, corticated and foliose functional-morphological groups were dominant at 1.5 m depth, the articulated calcareous, encrusting and foliose ones at 5 m depth and the encrusting and filamentous ones at 25 m depth. Due to the strong exposure, the leathery macroalgal group was poorly developed [55,72], and represented only by *C. compressa*, with very low abundance. From a faunal point of view the present data on polychaetes differ considerably from previous studies, with 23 species recently reported for the first time in the North Adriatic region, including four new records for the Adriatic Sea, as described elsewhere [43,44,73], while the sabellid *Amphicorina rovignensis*, collected at Sveti Ivan during our survey, was recently described as new to science [44]. Peculiarly, some of the newly reported species were among the most abundant ones in the examined habitats, in particular *Nereis usticensis, Syllis corallicola, S. gerlachi, S. rosea* and, the most frequent species in this research, *S. pirifera*. Results of faunal and biogeographic analyses of Syllidae from Sveti Ivan Island indicated possible "meridionalization" of the North Adriatic fauna, i.e., the process of establishment of native Mediterranean warm-water species typical from southern sectors in colder northern areas [43,74]. Two species recorded during our survey are considered alien in the Mediterranean being considered Lessepsian migrants, *Nereis persica* Fauvel, 1911 and *Lepidonotus tenuisetosus* (Gravier, 1902) [73,75] and are deposited in the collection of the Natural History Museum in Rijeka (Croatia). *Nereis persica* was previously reported in the Adriatic Sea only twice, in its northern part [76,77]; however, those records were considered questionable [75]. Elsewhere in the Mediterranean, it was reported along the coast of Israel and Turkey [78,79]. If previous records of this species in the Adriatic Sea are eventually erroneous [75], our finding could represent a north-western widening of the species areal in the Mediterranean. *Lepidonotus tenuisetosus* was until now reported only in the Eastern Mediterranean, along the coasts of Israel, Egypt, Turkey and Greece [75,79–81]. Our finding of *L. tenuisetosus* represents the first report for the Adriatic Sea and might indicate a north-western widening of the species' distribution. However, present data are not fully comparable with previous studies from a qualitative point of view, because the knowledge on polychaetes taxonomy is continuously evolving and new species are recorded and newly described also in the Adriatic Sea [73]. New records of some species in our research (such as *Nereis usticensis* Cantone, Catalano & Badalamenti, 2001 and several syllid species) might arise from the fact that former studies were carried out before those species were scientifically described [43]. Moreover, polychaete systematics changed a lot meanwhile and some taxa are not considered valid at present, such as for example previously reported *Pionosyllis serrata* Southern, 1914, *Syllis brevipennis* (Grube, 1863) and *Vermiliopsis richardi* Fauvel, 1909 [33,37], herein listed as *Nudisyllis pulligera* (Krohn, 1852), *Trypanosyllis (Trypanosyllis) coeliaca* Claparède, 1868 and *Vermiliopsis labiata*

(O. G. Costa, 1861) respectively. All this emphasizes the importance of research such as the present one, aiming at updating the knowledge of polychaetes biodiversity of an area.

Our data are more comparable to more recent studies conducted with similar experimental designs in the South Adriatic Sea by Giangrande et al. [82,83]. These authors found 152 polychaete taxa from 22 families in the first study [82] and 118 taxa from 18 families in the second one [83]. Species mostly characterising 5 m and 25 m depth in the south Adriatic [82] were mainly different from those characterising the same depths in our study, with only *P. dumerilii* being in common at 5 m and *S. armillaris* and *S. gerlachi* at 25 m depth. These differences could derive from local differences in the composition of algal assemblages, from different sampling periods (May and November in Giangrande et al. [82] and July in our research) and also from the biogeographic distribution of the Adriatic Sea polychaete fauna. Several species that characterised 5 m and 25 m depth in the south Adriatic were overall absent in our research, in particular *Syllis pulvinata* (Langerhans, 1881) that characterised 5 m and 25 m depth and *S. golfonovensis* (Hartmann-Schröder, 1962) and *Kefersteinia cirrhata* (Keferstein, 1862) that characterised 25 m depth. In fact, *S. pulvinata* and *S. golfonovensis* are species that were up to date found only in the southern part of the Adriatic, while it remains difficult to explain the absence of *K. cirrata* in our samples, since this species is reported as widely distributed in the whole Adriatic Sea [73]. This species is reported as *Psamathe fusca* Johnston, 1836 in the recent-most polychaete checklist of the Adriatic Sea, based on the synonymy proposed by Pleijel [84]. Although high, the number of taxa from Sveti Ivan Island is considerably lower than that reported in Giangrande et al. [82], possibly because of the smaller sampling surface and less extensive sampling period. Indeed, species area estimator curves suggested potentially higher species richness in our study area. Many species were rare and the analyses of distribution patterns indicating that additional sampling would probably yield more species, as well as it would presumably do sampling in different seasons, considering potential seasonal variability of hard bottom polychaete assemblages [51,58]. Thus, further studies encompassing different seasons and spatial scales should be done in order to upgrade the knowledge of diversity and spatial-temporal variation of hard bottom polychaete assemblages in the North Adriatic region.

The most abundant and species rich families found herein (particularly the Syllidae, but also the Sabellidae, Nereididae, Eunicidae and Serpulidae) are commonly reported as the most characteristic in the Mediterranean hard bottom polychaete assemblages [52,58,82,85,86].

The structure of the polychaete assemblages was highly variable alongshore, but only at the shallowest sites, while it clearly varied bathymetrically, which agrees with the most common trends previously identified, e.g., [11,82]. High variability in the shallowest habitats could be promoted by the high variability of environmental factors (temperature, salinity, hydrodynamics, light intensity etc.) [45]. However, the small sample size used herein might account for differences in presence/absence of several taxa that were represented by few individuals in the overall samples analysed and might have influenced the observed patterns of variation. The highest abundance and diversity and the highest similarity among polychaete assemblages was found at 5 m depth, as well as the highest number of species contributing to the similarity between samples, indicating that assemblages at this depth are the most diverse and structurally complex [87]. The high complexity of the algal coverage at intermediate depth, which can be considered an ecotone where photophilic and sciaphilic conditions coexist, together with a decrease in environmental variability, compared to the shallower habitat, possibly explain the observed increase in diversity [83,88]. In fact, at 5 m depth, the rich coverage of foliose algae (*Padina pavonica*, *Flabelia petiolata* and *Dictyota dichotoma*), structurally complex articulated calcareous algae (*Corallina officinalis* and *Jania* sp.) and encrusting algae (*P. rubra*, *P. heteromorpha*), with the last two forms being known to entrap considerable quantities of sediment, created altogether a complex and heterogeneous habitat suitable for diverse epifaunal and infaunal polychaete species. We expected the assemblages from 25 m depth to show the highest homogeneity as a consequence of the supposedly more stable environmental conditions. However, the within group similarity was lower at 25 m depth than at 5 m depth, likely because the sites at 25 m depth were situated at the end of the rocky slope, very close to the soft bottom, which may give rise to occasional sedimentation. Enhanced sedimentation combined with reduced light intensity, together with the simplification of the algal-habitat structure, could also contribute to explaining the lower abundance and species richness at 25m depth [48,89]. However, we cannot exclude an effect of the limited sampling effort in the observed patterns of polychaete assemblage variation.

There was a common trend along the whole research area, characterised by differences in the structure of polychaete assemblages between the shallower depths (1.5 m and 5 m) and 25 m depth as already reported elsewhere [82]. The assemblages from 1.5 m and 5 m depths showed significant differences in their structure only at one station. In fact, assemblages from these shallower habitats were commonly characterised by species typical of shallow photophilic environments reported within a variety of algal assemblages (such are *Amphiglena mediterranea*, *Platynereis dumerilii*, *Syllis prolifera* and *Exogone dispar*) [52,53,90,91], while those from 25 m depth were characterised by species usually also found in sciaphilic habitats (such are *S. armillaris* and *S. gerlachi*) [82,90]. The increase of species richness of Serpulidae with depth observed in our research was already reported elsewhere in the Mediterranean [57,92]. It is probably related to the combining effects of the increment of hard (both lithic and organogenic) substrata, low hydrodynamic energy and shadowing, which favour the development of underlying biogenic concretions hosting species with coralligenous affinity [92]. As a whole, the distribution pattern of polychaete assemblages in the examined area appeared related mostly to depth, which, per se, covaries with different environmental variables (temperature, light intensity, hydrodynamics, sedimentation, etc). It was also shown to be related to the algal distribution, being correlation higher considering single algal species than algal functional-morphological groups. The encrusting calcareous alga *P. rubra* and the articulated calcareous alga *Haliption* sp. ranked first among algal predictor variables of polychaete distribution, followed by one filamentous and two foliose species. The high percent coverage of articulated calcareous and encrusting algae certainly contributed to explain the high abundance and diversity of polychaetes, due to their high structural complexity that provided wide panoply of suitable microhabitats [88,93]. However, taxa with less complex morphology, i.e., the unidentified filamentous species, also contributed to significantly explain the variability of the polychaete assemblages. In particular, filamentous algal species characterised the habitat at 25 m depth. The spatial complexity of algal thalli surely represents an important factor influencing polychaete distribution, but other non-three-dimensional algal features (e.g., production of antagonistic metabolites, epiphyte colonization, palatability, capability to entrap sediment, etc.) could also be important and deserve further investigations [51,94].

In recent years, the North Adriatic macroalgal assemblages have suffered severe changes (such as reduction in algal coverage, variation in richness and species composition and simplification of the community structure), due to different natural and human driven impacts [95,96], which could likely provoke alterations of the resident polychaete assemblages. However, our algal-based predictor variables explained only part of the observed variability of the polychaete assemblages. We assume that other environmental variables, as well as biotic interactions among polychaetes and between them and other benthic invertebrates (e.g., competition for food or space, recruitment, predation, etc.) may contribute to explain part of the unexplained variability and, thus, should be considered in future studies [11,51,82].

Our results suggest that the diversity of the North Adriatic hard bottom polychaete fauna may be largely underestimated. Further faunal and ecological studies over larger spatial and temporal scale are thus welcome in order to implement our knowledge on diversity and distribution of polychaete assemblages in the area, which will serve as a necessary base to detect changes and predict consequences of natural and anthropogenic disturbances on benthic communities in this important and sensitive Mediterranean sector.

Supplementary Materials: The following are available online at http://www.mdpi.com/1424-2818/12/10/408/s1, Table S1: Mean coverage (MC) (%) (± standard error (±SE), in italics) of algal species and functional-morphological groups per station and depth. Most characteristic species are marked with asterisk (*); Table S2: Mean abundance (MA) (± standard error (±SE), in italics) of Polychaete taxa per station and depth. In bold values of most abundant taxa on each station and depth; Table S3. Results of PERMANOVA pair-wise and PERMDISP analyses testing for differences in abundance (N), species richness (S), Hill's species diversity index (N1), Hill's evenness index (N10) and structure (Stru) of polychaete assemblages: between stations, separately for each depth and between depths, separately for each station. Up, unique perms; t, t-test; P (perm), probability; P (MC), Monte Carlo probability. Significant P-values ($p < 0.05$) are given in bold; Table S4. Results of SIMPER analyses (cut-off 70%) used to identify taxa that mostly contribute to faunal dissimilarity between stations at each depth. Species contributing to dissimilarity with more than 4% are marked with asterisk. Av.Ab = mean abundance, Diss = mean dissimilarity, Diss/SD = dissimilarity/standard deviation, Contrib% = contribution relative to single taxon, Cum% = cumulative contribution; Figure S1. Distribution of species according to their frequency in the studied samples; Figure S2. Species area accumulation curve (Sobs, Species observed) and estimator curves (Jacknife1, Jacknife2 and Bootstrap).

Author Contributions: The study was conceived and designed by B.M., A.G. and L.M., B.M. and A.J. conducted sampling campaign. Analyses of benthic assemblages and polychaetes were done by B.M., A.G., M.L. and L.M., while Lj.I. analysed algae. Data analysis was performed by B.M. and L.M. The first draft of the manuscript was written by B.M. and all authors commented on previous versions of the manuscript. All authors have read and agreed to the published version of the manuscript.

Funding: This research was funded by THE GOVERNMENT OF THE REPUBLIC OF ITALY through a research grant for Croatian citizens for the academic year 2007/2008, awarded to the author BM for the execution of the project "Faunal and biogeographic analyses of the Adriatic Sea polychaetes: taxonomic comparison of the polychaetes of Istrian and Salento coasts".

Conflicts of Interest: The authors declare no conflict of interest. The funders had no role in the design of the study; in the collection, analyses, or interpretation of data; in the writing of the manuscript, or in the decision to publish the results.

References

1. Buljan, M.; Zore-Armanda, M. Oceanographical properties of the Adriatic Sea. *Oceanogr. Mar. Biol. Annu. Rev.* **1976**, *14*, 11–98.
2. McKinney, F.K. *The Northern Adriatic Ecosystem: Deep Time in a Shallow Sea*; Columbia University Press: New York, NY, USA, 2007; 328p.
3. Degobbis, D.; Gilmartin, M. Nitrogen, phosphorus and biogenic silicon budgets for the northern Adriatic Sea. *Oceanol. Acta* **1990**, *13*, 31–45.
4. Boero, F.; Bonsdorff, E. A conceptual framework for marine biodiversity and ecosystem functioning. *Mar. Ecol Evol Perspect* **2007**, *28* (Suppl. 1), 134–145. [CrossRef]
5. Mikac, B.; Semprucci, F.; Guidi, L.; Ponti, M.; Abbiati, M.; Balsamo, M.; Dovgal, I. Newly discovered associations between peritrich ciliates (Ciliophora: Peritrichia) and scale polychaetes (Annelida: Polynoidae and Sigalionidae) with a review of polychaete–peritrich epibiosis. *Zool. J. Linn. Soc* **2019**, *20*, 1–15. [CrossRef]
6. Coll, M.; Piroddi, C.; Albouy, C.; Lasram, F.B.R.; Cheung, W.W.L.; Christensen, V.; Karpouzi, V.S.; Guilhaumon, F.; Mouillot, D.; Paleczny, M.; et al. The Mediterranean Sea under siege: Spatial overlap between marine biodiversity, cumulative threats and marine reserves. *Glob. Ecol. Biogeogr.* **2012**, *21*, 465–480. [CrossRef]
7. Coll, M.; Piroddi, C.; Steenbeek, J.; Kaschner, K.; Ben Rais Lasram, F.; Aguzzi, J.; Ballesteros, E.; Bianchi, C.N.; Corbera, J.; Dailianis, T.; et al. The Biodiversity of the Mediterranean Sea: Estimates, patterns, and threats. *PLoS ONE* **2012**, *5*. [CrossRef]
8. Micheli, F.; Halpern, B.S.; Walbridge, S.; Ciriaco, S.; Ferretti, F.; Fraschetti, S.; Lewison, R.; Nykjaer, L.; Rosenberg, A.A. Cumulative Human Impacts on Mediterranean and Black Sea Marine Ecosystems: Assessing Current Pressures and Opportunities. *PLoS ONE* **2013**, *8*, e79889. [CrossRef]
9. European Commission. Directive 2008/56/EC of the European Parliament and of the Council of 17 June 2008 establishing a framework for Community action in the field of marine environmental policy (Marine Strategy Framework Directive). *Off. J. Eur. Union* **2008**, *L 164, 25.6.2008*, 19–40.
10. Knox, G.A. The role of Polychaetes in benthic soft-bottom communities. In *Essays on Polychaetous Annelids in Memory of Olga Hartmann*; Reish, D., Fauchald, K., Eds.; Allan Hancock Foundation: Los Angeles, CA, USA, 1977; pp. 547–604.

11. Musco, L. Ecology and diversity of Mediterranean hard-bottom Syllidae (Annelida): A community-level approach. *Mar. Ecol Prog. Ser.* **2012**, *461*, 107–119. [CrossRef]

12. Giangrande, A.; Licciano, M.; Musco, L. Polychaetes as environmental indicators revisited. *Mar. Pollut Bull.* **2005**, *50*, 1153–1162. [CrossRef]

13. Aguado-Giménez, F.; Gairin, J.I.; Martinez-Garcia, E.; Fernandez-Gonzalez, V.; Ballester Molto, M.; Cerezo-Valverde, J.; Sanchez-Jerez, P. Application of taxocene surrogation and taxonomic sufficiency concepts to fish farming environmental monitoring. Comparison of BOPA index versus polychaete assemblage structure. *Mar. Environ. Res.* **2015**, *103*, 27–35. [CrossRef] [PubMed]

14. Musco, L.; Terlizzi, A.; Licciano, M.; Giangrande, A. Taxonomic structure and the effectiveness of surrogates in environmental monitoring: A lesson from polychaetes. *Mar. Ecol Prog. Ser.* **2009**, *383*, 199–210. [CrossRef]

15. Soares-Gomes, A.; Mendes, C.L.T.; Tavares, M.; Santi, L. Taxonomic sufficiency of polychaete taxocenes for estuary monitoring. *Ecol Indic.* **2012**, *15*, 149–156. [CrossRef]

16. Aleffi, F.; Bettoso, N.; Solis-Weiss, V. Spatial distribution of soft–bottom polychaetes along western coast of the northern Adriatic Sea (Italy). *Ann. Ser. Hist Nat.* **2003**, *13*, 211–222.

17. Amoureux, L. Inventaire d'une petit collection d'Annélides Polychètes des parages sud de Rovinj (Haute–Adriatique). *Thalass. Jugosl.* **1976**, *12*, 381–390.

18. Amoureux, L. Annélides Polychètes recueillies par D. Zavodnik. *Thalass. Jugosl.* **1983**, *19*, 1–6.

19. Castelli, A.; Prevedelli, D. Effetto del fenomeno delle mucillaggini dell'estate 1989 sul popolamento a policheti di un microhabitat salmastro preso Punta Marina (Ravenna). *Biol. Mar. Meditsuppl. Al Not. S.I.B.M.* **1993**, *1*, 35–38.

20. Crema, R.; Prevedelli, D.; Valentini, A.; Castelli, A. Recovery of the macrozoobenthic community of the Comacchio lagoon system (Northern Adriatic Sea). *Ophelia* **2000**, *52*, 143–152. [CrossRef]

21. Fauvel, P. Annélides Polychètes de Rovigno d'Istria. *Thalassia* **1934**, *1*, 1–78.

22. Fauvel, P. Annelida Polychaeta della Laguna di Venezia. *Bollettino. R. Com. Talassogr. Ital.* **1938**, *246*, 1–27.

23. Fauvel, P. Annélides Polychètes de la Haute Adriatique. *Thalassia* **1940**, *4*, 1–24.

24. Gamulin-Brida, H.; Požar, A.; Zavodnik, D. Contributions aux recherches sur la bionomie benthique des fonds meubles de l'Adriatique du nord (II). *Biol. Glas* **1968**, *21*, 157–201.

25. Gillet, P. Annélides Polychètes des fonds meubles du Canal de Lim prés de Rovinj (Yugoslavie). *Thalassia Jugosl.* **1986**, *22/21*, 127–138.

26. Maggiore, F.; Keppel, E. Biodiversity and distribution of polychaetes and molluscs along the Dese estuary (Lagoon of Venice, Italy). *Hydrobiologia* **2007**, *588*, 189–203. [CrossRef]

27. Mikac, B.; Musco, L.; Đakovac, T.; Giangrande, A.; Terlizzi, T. Long–term changes in North Adriatic soft–bottom polychaete assemblages following a dystrophic crisis. *Ital. J. Zool.* **2011**, *78*, 304–316. [CrossRef]

28. Mistri, M.; Fano, A.E.; Ghion, F.; Rossi, R. Disturbance and community pattern of polychaetes inhabiting Valle Magnavacca (Valli di Comacchio, Northern Adriatic Sea, Italy). *Mar. Ecol. Pszni* **2002**, *23*, 31–49. [CrossRef]

29. Occhipinti-Ambrogi, A.; Savini, D.; Forni, G. Macrobenthos community structural changes off Cesenatico coast (Emilia Romagna, Northern Adriatic), a six–year monitoring programme. *Sci. Tot Environ.* **2005**, *353*, 317–328. [CrossRef]

30. Prevedelli, D.; Bellucci, L.G.; Simononi, R.; Ansaloni, I.; Frignani, M.; Ravaioli, M.; Castelli, A. Macrobenthos and environmental characteristics of the Venice lagoon. *Atti. Soc. Nat. Mat. Modena* **2007**, *138*, 151–161.

31. Zahtila, E. Offshore polychaete fauna in the northern Adriatic with trophic characteristic. *Period. Biol.* **1997**, *99*, 213–217.

32. Zavodnik, D.; Vidaković, J.; Amoureux, L. Contribution to sediment macrofauna in the area of Rovinj (North Adriatic Sea). *Cah Biol. Mar.* **1985**, *26*, 431–444.

33. Amoureux, L. Annélides Polychètes de l'îlot Banjole (prés de Rovinj, haute–Adriatique). *Cah Biol. Mar.* **1975**, *16*, 231–244.

34. Amoureux, L.; Katzmann, W. Note faunistique et écologique sur une collection d'Annélides Polychètes de substrats rocheux circalittoraux de la région de Rovinj (Yougoslavie). *Zool Anz* **1971**, *186*, 114–122.

35. Casellato, S.; Masiero, L.; Sichirollo, E.; Soresi, S. Hidden secrets of the Northern Adriatic: "Tegnue", peculiar reefs. *Centr. Eur. J. Biol.* **2007**, *2*, 122–136. [CrossRef]

36. Casellato, S.; Stefanon, A. Coralligenous habitat in the northern Adriatic Sea: An overview. *Mar. Ecol.* **2008**, *29*, 321–341. [CrossRef]

37. Katzmann, W. Polychaeten (Errantier, Sedentarier) aus nord—Adriatischen *Cystoseira*—Beständen und deren Epiphyten. *Oecologia* **1971**, *8*, 31–51. [CrossRef] [PubMed]

38. Pitacco, V.; Mavrič, B.; Orlando-Bonaca, M.; Lipej, L. Rocky macrozoobenthos mediolittoral community in the Gulf of Trieste (North Adriatic) along a gradient of hydromorphological modifications. *Acta Adriat.* **2013**, *54*, 67–86.

39. Pitacco, V.; Orlando-Bonaca, M.; Mavrič, B.; Lipej, L. Macrofauna associated with a bank of *Cladocora caespitosa* (Anthozoa, scleractinia) in the Gulf of Trieste (northern Adriatic). *Ann. Ser. Hist Nat.* **2014**, *24*, 1–14.

40. Zavodnik, D. Prispevk k poznavanju naselja *Cystoseira barbata* (Good. & Wood.) C. Ag. v severnem Jadranu. *Biol. Vest* **1965**, *13*, 87–101.

41. Zavodnik, D. Dinamika litoralnega fitala na zahodnoistrski obali. *Razpr. Slov. Akad. Znan. Umet.* **1967**, *10*, 5–71.

42. Zavodnik, D.; Legac, M.; Gluhak, T. An account of the marine fauna of Pag Island (Adriatic Sea, Croatia). *Nat. Croat.* **2006**, *15*, 65–107.

43. Mikac, B.; Musco, L. Faunal and biogeographic analysis of Syllidae (Polychaeta) from Rovinj (Croatia, northern Adriatic Sea). *Sci. Mar.* **2010**, *74*, 353–370. [CrossRef]

44. Mikac, B.; Liciano, M.; Giangrande, A. Sabellidae and Fabriciidae (Polychaeta) of the Adriatic Sea with particular retrospect to the Northern Adriatic and the description of two new species. *J. Mar. Biol. Assoc. Uk* **2013**, *93*, 1511–1524. [CrossRef]

45. Benedetti-Cecchi, L.; Bulleri, F.; Cinelli, F. The interplay of physical and biological factors in maintaining midshore and low-shore assemblages on rocky coasts in the north-west Mediterranean. *Oecologia* **2000**, *123*, 406–417. [CrossRef] [PubMed]

46. Çinar, M.E. Ecological features of Syllidae (Polychaeta) from shallow-water benthic environments of the Aegean Sea, eastern Mediterranean. *J. Mar. Biol. Ass Uk* **2003**, *83*, 737–745. [CrossRef]

47. Fraschetti, S.; Terlizzi, A.; Benedetti-Cecchi, L. Patterns of distribution of rocky marine assemblages: Evidence of relevant scales of variation. *Mar. Ecol. Prog. Ser.* **2005**, *296*, 13–29. [CrossRef]

48. Irving, A.D.; Connell, S.D. Sedimentation and light penetration interact to maintain heterogeneity of subtidal habitats: Algal versus invertebrate dominated assemblages. *Mar. Ecol. Prog. Ser.* **2002**, *245*, 83–91. [CrossRef]

49. Terlizzi, A.; Anderson, M.J.; Fraschetti, S.; Benedetti-Cecchi, L. Scales of spatial variation in Mediterranean subtidal sessile assemblages at different depths. *Mar. Ecol. Prog. Ser.* **2007**, *332*, 25–39. [CrossRef]

50. Underwood, A.J.; Chapman, M.G. Scales of spatial patterns of distribution of intertidal invertebrates. *Oecologia* **1996**, *107*, 212–224. [CrossRef]

51. Dorgham, M.M.; Hamdy, R.; El-Rashidy, H.H.; Atta, M.M.; Musco, L. Distribution patterns of shallow water polychaetes (Annelida) along the coast of Alexandria, Egypt (eastern Mediterranean). *Mediterr. Mar. Sci.* **2014**, *15*, 635–649. [CrossRef]

52. Giangrande, A. Polychaete zonation and its relation to algal distribution down a vertical cliff in the western Mediterranean (Italy): A structural analysis. *J. Exp. Mar. Biol. Ecol.* **1988**, *120*, 263–276. [CrossRef]

53. Somaschini, A. Policheti della biocenosi ad alghe fotofile (Facies a *Corallina elongata*) nel Lazio settentrionale. *Atti Sac. Toscana Sci. Nat. Mem Ser. B* **1988**, *95*, 83–94.

54. Iveša, L. Dinamika Popluacija Makrofitobentosa na Hridinastim Dnima uz Zapadnu Obalu Istre. Ph.D. Thesis, Faculty of Science, University of Zagreb, Zagreb, Croatia, 2005.

55. Iveša, Lj.; Lyons, D.M.; Devescovi, M. Assessment of the ecological status of north-eastern Adriatic coastal waters (Istria, Croatia) using macroalgal assemblages for the European Union Water Framework Directive. *Aquat. Conserv. Mar. Freshw. Ecosyst.* **2009**, *19*, 14–23. [CrossRef]

56. Boudouresque, C. Méthodes d'étude qualitative et quantitative du benthos (en particulier du phytobenthos). *Tethys* **1971**, *3*, 79–104.

57. Casoli, E.; Bonifazi, A.; Ardizzone, G.; Gravina, M.F. How algae influence sessile marine organisms: The tubeworms case of study. *Estuar. Coast. Shelf Sci.* **2016**, *178*, 12–20. [CrossRef]

58. Fraschetti, S.; Giangrande, A.; Terlizzi, A.; Miglietta, M.P.; Della Tommasa, L.; Boero, F. Spatio-temporal variation of hydroids and polychaetes associated with *Cystoseira amentacea* (Fucales: Phaeophyceae). *Mar. Biol.* **2002**, *140*, 949–957. [CrossRef]

59. Cormaci, M.; Furnari, G.; Giaccone, G. Macrophytobenthos. In *Mediterranean Marine Benthos: A Manual of Methods for Its Sampling and Study*; Gambi, M.C., Dappiano, M., Eds.; Biol. Mar. Medit. 11 (1), Chapter 7; Società Italiana di Biologia Marina: Genova, Italy, 2004; pp. 217–246.

60. Littler, M.M.; Littler, D.S. The evolution of thallus form and survival strateges in benthic marine macroalgae: Field and laboratory tests of a functional form model. *Am. Nat.* **1980**, *116*, 5–44. [CrossRef]

61. Littler, M.M.; Littler, D.S. Relationships between macroalgal functional form groups and substrata stability in a subtropical rocky-intertidal system. *J. Exp. Mar. Biol. Ecol.* **1984**, *74*, 13–34. [CrossRef]
62. Magurran, A.E. *Measuring Biological Diversity*; Blackwell Publishing: Oxford, UK, 2004; 256p.
63. Colwell, R.K.; Coddington, J.A. Estimating terrestrial biodiversity through extrapolation. In *Biodiversity: Measurement and Estimation*; Hawksworth, D.L., Ed.; The Royal Society: London, UK, 1994; pp. 101–118.
64. Anderson, M.J. A new method for non-parametric multivariate analysis of variance. *Aust. J. Ecol.* **2001**, *26*, 32–46. [CrossRef]
65. McArdle, B.H.; Anderson, M.J. Fitting multivariate models to community data: A comment on distance-based redundancy analysis. *Ecology* **2001**, *82*, 290–297. [CrossRef]
66. Anderson, M.J.; Gorley, R.N.; Clarke, K.R. *PERMANOVA+ for PRIMER: Guide to Software and Statistical Methods*; PRIMER-E: Plymouth, UK, 2008.
67. Clarke, K.R. Nonparametric multivariate analyses of changes in community structure. *Aust. J. Ecol.* **1993**, *18*, 17–143. [CrossRef]
68. Anderson, M.J. *DISTLM Forward: A FORTRAN Computer Program to Calculate a Distance-Based Multivariate Analysis for a Linear Model Using Forward Selection*; Department of Statistics, University of Auckland: Auckland, New Zealand, 2003.
69. Clarke, K.R.; Gorley, R.N. *PRIMER v6: User Manual/Tutorial*; PRIMER-E: Plymouth, UK, 2006; 192p.
70. Pérès, J.M.; Gamulin-Brida, H. *Biološka Oceanografija. Bentos. Bentoska Bionomija Jadranskog Mora*; Školska knjiga: Zagreb, Croatia, 1973; 493p.
71. Orfanidis, S.; Panayotidis, P.; Stamatis, N.V. An insight to ecological evaluation index (EEI). *Ecol. Indic.* **2003**, *3*, 27–33. [CrossRef]
72. Munda, I.M. Changes in the benthic algal associations of the vicinity of Rovinj (Istrian coast, North Adriatic) caused by organic wastes. *Acta Adriat* **1980**, *21*, 299–332.
73. Mikac, B. A sea of worms: Polychaete checklist of the Adriatic Sea. *Zootaxa* **2015**, *3943*, 1–172. [CrossRef] [PubMed]
74. Bianchi, C.N. Biodiversity issues for the forthcoming tropical Mediterranean Sea. *Hydrobiologia* **2007**, *580*, 7–21. [CrossRef]
75. Zenetos, A.; Gofas, S.; Verlaque, M.; Çinar, M.E.; García Raso, J.E.; Bianchi, C.N.; Morri, C.; Azzurro, E.; Bilecenoglu, M.; Froglia, C.; et al. Alien species in the Mediterranean Sea by 2010. A contribution to the application of European Union's Marine Strategy Framework Directive (MSFD). Part I. Spatial distribution. *Medit Mar. Sci.* **2010**, *11/2*, 381–493. [CrossRef]
76. Amoureux, L. Annélides Polychètes de l'Adriatique. *Thalass. Jugosl.* **1983**, *19*, 15–21.
77. Zavodnik, D.; Kovačić, M. Index of marine fauna in Rijeka Bay (Adriatic Sea, Croatia). *Nat. Croat.* **2000**, *9*, 297–379.
78. Ben-Eliahu, M.N. Polychaeta errantia of the Suez Canal. *Isr. J. Zool.* **1972**, *21*, 189–237.
79. Çinar, M.E. Alien polychaete species (Annelida: Polychaeta) on the southern coast of Turkey (Levantine Sea, eastern Mediterranean), with 13 new records for the Mediterranean Sea. *J. Nat. Hist* **2009**, *43*, 2283–2328. [CrossRef]
80. Barnich, R.; Fiege, D. The Aphroditoidea (Annelida: Polychaeta) of the Mediterranean Sea. *Abh. Der. Senckenbergischen Nat. Ges. Frankfut Am. Main* **2003**, *559*, 1–170.
81. Chatzigeorgiou, G.; Faulwetter, S. Polychaetes from Two Subtidal Rocky Shores of the North Coast of Crete, Collected for the NaGISA Project 2007–2008. Version 1.8, 2020, Sampling Event Dataset. Available online: https://www.gbif.org/occurrence/1705514079 (accessed on 13 April 2020). [CrossRef]
82. Giangrande, A.; Delos, A.L.; Fraschetti, S.; Musco, L.; Licciano, M.; Terlizzi, A. Polychaete assemblages along a rocky shore on the South Adriatic coast (Mediterranean Sea): Patterns of spatial distribution. *Mar. Biol.* **2003**, *143*, 1109–1116. [CrossRef]
83. Giangrande, A.; Delos, A.L.; Musco, L.; Licciano, M.; Pierri, C. Polychaete assemblages of rocky shore along the South Adriatic coast (Mediterranean Sea). *Cah Biol. Mar.* **2004**, *45*, 85–95.
84. Pleijel, F. Phylogeny and classification of Hesionidae (Polychaeta). *Zool Scr.* **1998**, *27*, 89–163. [CrossRef]
85. Çinar, M.E.; Gönlügür-Demirci, G. Polychaete assemblages on shallow-water benthic habitats along the Sinop Peninsula (Black Sea, Turkey). *Cah Biol. Mar.* **2005**, *46*, 253–263.

86. Gambi, M.C.; Musco, L.; Giangrande, A.; Badalamenti, F.; Micheli, F.; Kroeker, K.J. Distribution and functional traits of polychaetes in a CO_2 vent system: Winners and losers among closely related species. *Mar. Ecol. Prog. Ser.* **2016**, *550*, 121–134. [CrossRef]

87. Dahl, L.; Dahl, K. Temporal, spatial and substrate-dependent variations of Danish hard-bottom macrofauna. *Helgol. Mar. Res.* **2002**, *56*, 159–168. [CrossRef]

88. Serrano, A.; Preciado, I. Environmental factors structuring polychaete communities in shallow rocky habitats: Role of physical stress versus habitat complexity. *Helgol Mar. Res.* **2007**, *61*, 17–29. [CrossRef]

89. Airoldi, L. The effects of sedimentation *on* rocky coast assemblages. *Oceanogr Mar. Biol. Ann. Rev.* **2003**, *41*, 161–236.

90. Sardá, R. Polychaete communities related to plant covering in the midlittoral and infralittoral zones of the Balearic Islands (Western Mediterranean). *Pszni Mar. Ecol.* **1991**, *12*, 341–360. [CrossRef]

91. López, E.; Viéitez, J.M. Polychaete assemblages on non-encrusting infralittoral algae from the Chafarinas Islands (SW Mediterranean). *Cah Biol. Mar.* **1999**, *40*, 375–384.

92. Sanfilippo, R.; Rosso, A.; Sciuto, F.; Serio, D.; Catra, M.; Alongi, G.; Negri, M.P.; Leonardi, R.; Viola, A. Serpulid polychaetes from Cystoseira communities in the Ionian Sea, Mediterranean. *Vie Et Milieu* **2017**, *67*, 217–226.

93. Tena, J.; Capaccioni-Azzati, R.; Torres-Gravila, F.J.; Garcia-Carrascosa, A.M. Polychaetes associated with different facies of the photophilic algal community in the Chafarinas archipelago (SW Mediterranean). *Bull. Mar. Sci.* **2000**, *67*, 55–72.

94. Cacabelos, E.; Olabarria, C.; Incera, M.; Troncoso, J.S. Effects of habitat structure and tidal height on epifaunal assemblages associated with macroalgae. *Estuar Coast. Shelf Sci.* **2010**, *89*, 43–52. [CrossRef]

95. Falace, A.; Alongi, G.; Cormaci, M.; Furnari, G.; Curiel, D.; Cecere, E.; Petrocelli, A. Changes in the benthic algae along the Adriatic Sea in the last three decades. *Chem. Ecol.* **2010**, *26*, 77–90. [CrossRef]

96. Marchini, A.; Ragazzola, F.; Vasapollo, C.; Castelli, A.; Cerrati, G.; Gazzola, F.; Jiang, C.; Langeneck, J.; Manauzzi, M.C.; Musco, L.; et al. Intertidal Mediterranean Coralline Algae Habitat Is Expecting a Shift Toward a Reduced Growth and a Simplified Associated Fauna Under Climate Change. *Front. Mar. Sci.* **2019**, *6*, 106. [CrossRef]

 *diversity*

Article

Filling the Gap and Improving Conservation: How IUCN Red Lists and Historical Scientific Data Can Shed More Light on Threatened Sharks in the Italian Seas

Francesco Luigi Leonetti [1,2], Emilio Sperone [1], Andrea Travaglini [3], Angelo R. Mojetta [4], Marco Signore [3], Peter N. Psomadakis [5], Thaya M. Dinkel [2] and Massimiliano Bottaro [2,*]

[1] Department of Biology, Ecology and Earth Sciences (DiBEST), University of Calabria, 87036 Arcavacata di Rende (CS), Italy; francescoluigi.leonetti@szn.it (F.L.L.); emilio.sperone@unical.it (E.S.)

[2] Department of Integrative Marine Ecology (EMI), Stazione Zoologica Anton Dohrn, Italian National Institute for Marine Biology, Ecology and Biotechnology, Villa Comunale, 80121 Naples, Italy; thaya.dinkel@szn.it

[3] Darwin Dohrn Museum (DaDoM), Stazione Zoologica Anton Dohrn, Italian National Institute for Marine Biology, Ecology and Biotechnology, Villa Comunale, 80121 Naples, Italy; andrea.travaglini@szn.it (A.T.); marco.signore@szn.it (M.S.)

[4] Institute for Marine Studies (ISM), c/o Aquarium and Civic Hydrobiological Station of Milan, Viale Gadio, 2 20121 Milan, Italy; amojetta@tin.it

[5] Food and Agriculture Organization of the United Nations (FAO), Viale delle Terme di Caracalla, 00153 Rome, Italy; peter.psomadakis@gmail.com

* Correspondence: massimiliano.bottaro@szn.it; Tel.: +39-081-5833-280

Received: 9 September 2020; Accepted: 7 October 2020; Published: 10 October 2020

Abstract: Chondrichthyans are one of the most threatened marine taxa worldwide. This is also the case in the Mediterranean Sea, which is considered an extinction hotspot for rays and sharks. The central position of the Italian peninsula makes it an ideal location for studying the status and changes of this sea. There is a lack of biological, ecological and historical data when assessing shark populations, which is also highlighted in the Red List of Threatened Species compiled by the International Union for the Conservation of Nature (IUCN). Historical data can provide important information to better understand how chondrichthyan populations have changed over time. This study aims to provide a clearer understanding of the changes in distribution and abundance of eight shark species in the Italian seas that are currently classified as at risk of extinction by the IUCN. In this respect, a bibliographic review was conducted on items from the 19th century to the first half of the 20th century, focusing on the selected species. The results show that all sharks were considered common until the beginning of the 20th century but have declined since, with a clear negative trend, mainly in the past 70 years. The strong local decline has been attributed to overexploitation, bycatch, habitat loss, depletion of prey items and environmental pollution. Furthermore, historical data also allow us to avoid the issue of a 'shifting baseline', in which contemporary abundances are assumed to be "normal". Using historical data to further our knowledge of the marine environment is becoming increasingly common, and is fundamental in understanding human impact and evaluating mitigation measures to manage and conserve marine species and environments.

Keywords: biodiversity; chondrichthyans; conservation; fishing; historical ecology; Mediterranean Sea

---

1. Introduction

Chondrichthyes represents one of the most threatened marine taxa worldwide [1], and its species are highly susceptible to anthropogenic pressure, both in coastal and offshore environments [2,3].

In particular, the main factor affecting these vulnerable organisms is bycatch (incidental catch), which needs to be reduced to correctly manage and conserve cartilaginous fishes [4]. However, assessing and managing chondrichthyan populations is problematic, due to the limited nature of information on their biology and fisheries [5,6]. The absence of historical information about the population of cartilaginous fishes also makes it easy to fall pretty to "shifting baseline syndrome," in which one assumes that current conditions of resources are the standard, without taking into account their history of exploitation [7,8]. Furthermore, the few available long term studies are little more than snapshots, reflecting the vulnerability of the species under the fishing conditions they were subjected to when the data were collected [9]).

This lack of data is also highlighted in the List of Threatened Species (the Red List) of the International Union for the Conservation of Nature (IUCN). The Red List represents a comprehensive resource on the global status of biodiversity, and over the last decades it has become a broad and important tool for conservation, policy making and management. According to the IUCN, the Mediterranean Sea is one of the three areas in the world where the biodiversity of sharks and rays is most seriously threatened, with more than 40% of the assessed chondrichthyan species considered to be "endangered" (EN) or "critically endangered" (CR) [10]. Although the Mediterranean Sea represents a biodiversity hot spot [11], it also presents relevant geopolitical limits for the development of common and effective strategies of management and conservation of fisheries resources and natural heritage [12]. In this respect, the Italian peninsula is the core of the basin—it is placed on the natural border between the western and eastern sectors and it encompasses both the northern and southern sectors, functioning as a sort of "natural benchmark" of the general status and the changes in the Mediterranean Sea [13] and playing a potentially important role in the development of regional common initiatives for the management and protection of marine biodiversity [14,15].

Historical ecology developed as an organized research approach in the middle of the twentieth century [16]. It intersects strongly with environmental history, ecological anthropology, historical geography and paleoecology, with researchers from all disciplines contributing to the understanding of putative pristine environments and the anthropogenic changes they undergo [16–18]. Originally designed for landscape research, this scientific approach has been recently applied to the marine environment as well, focusing particularly on fishery resources [19–22]. From this point of view, even though the Mediterranean Sea is commonly regarded as the cradle of Western civilization [11], historical changes in its marine realm are less well understood [23,24], and historical research on cartilaginous fishes remains very limited and fragmented [25–30].

Considering the IUCN Red List of cartilaginous fishes in Italian waters [31], in this study we aim to provide novel scientific historical information from the last 250 years concerning the most threatened shark species, in order to contribute to a clearer understanding of their distribution before the modern fishery age, as well as the abundance and status of these chondrichthyans along the Italian peninsula, which is a key area for the marine biodiversity of the Mediterranean.

2. Materials and Methods

We selected ten shark species based on their risk of extinction from the Italian seas (Figure 1), according to the IUCN Red List of cartilaginous fishes in Italian waters [31]. In particular, we selected all species classified as critically endangered (CR), endangered (EN) and vulnerable (VU) shark species (Table 1).

Figure 1. Study area.

Table 1. List of threatened sharks in Italian seas considered in this study, based on the International Union for the Conservation of Nature (IUCN) Red List.

Species	Common Name	ItalianIUCN Red List (2012)	Mediterranean IUCN Red List (2016)
Squalus acanthias	spiny dogfish	CR	EN
Squatina aculeata	sawback angel shark	CR	CR
Squatina oculata	smoothback angel shark	CR	CR
Squatina squatina	angel shark	CR	CR
Alopias vulpinus	common thresher shark	CR	EN
Mustelus asterias	starry smoothhound	EN	VU
Mustelus mustelus	common smoothhound	EN	VU
Mustelus punctulatus	blackspotted smoothhound	EN	VU
Prionace glauca	blue shark	VU	CR
Galeorhinus galeus	tope shark	CR	VU

The research was conducted by analyzing scientific documents, research reports, papers and books on natural history from the 19th century through to the first half of the 20th century. Bibliographic items were researched via electronic archives (e.g., Google Scholar, ISI web of knowledge, national and/or local libraries) and through general web searches, using a combination of scientific, local and vernacular names of cartilaginous fishes in the different Italian maritime districts (Ligurian Sea, Tyrrhenian Sea, Ionian Sea and Adriatic Sea (Figure 1)). When available, records for the same species and time period were compared to ensure similar trends were observed. The records were classified by type, with a brief description for each item. All records were arranged in chronological order, by species, by citation and by geographical area.

All items were analyzed to ensure the inclusion of geographical information about fish abundance and interactions with humans. Particular attention was given to the frequency and abundance reported for the different species. This information was usually described in the historical bibliographic references using quantitative adjectives like "rare", "frequent", "common", etc. In order to parameterize the information, a progressive number system from one to six was associated with the quantitative terms, as summarized in Table 2. The categories considered by the IUCN (CR, EN, VU) were also parameterized in the same way (Table 2). The records are then discussed from both historical and ecological points of view, through comparison with the current status of Mediterranean chondrichthyans.

Table 2. Parameterization of the semiquantitative values of abundance recorded in the considered scientific references.

Reported Frequency/IUCN Category	Very Rare/CR	Rare/EN	Uncommon/VU	Frequent/NT	Common/LC	Very Common/LC
Assigned value	1	2	3	4	5	6

3. Results

We found scientific bibliographic references to eight of the ten selected shark species, with a temporal range from 1832 to 1962 (Table 3). All references were reported by scientists. The frequencies of species recorded in the different references over time are summarized in Figure 2. Based on the available information, semi-quantitative trends of abundance for the different species were assessed.

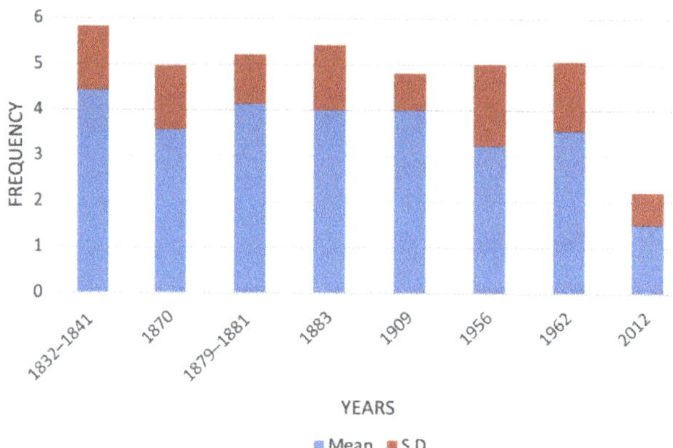

Figure 2. Frequency of species recorded in the different references.

Squalus acanthias was considered common along the Italian coast during the entire 19th century until the beginning of 1900 (Figure 3). Nevertheless, in the first decades after the Second World War (WWII) this species was considered uncommon, whereas today *S. acanthias* has been evaluated as very rare in the Italian seas and is listed as critically endangered (CR).

Squatina oculata was considered uncommon until the end of the 19th century when it was more commonly recorded (Figure 4). In the first decades after WWII it became rare again. Today, *S. oculata* is evaluated as very rare in the Italian seas and is listed as critically endangered (CR).

Table 3. References covering the selected shark species, with details on temporal range and related maritime sectors.

Species	Common name	Historical Periods, Related Maritime Sectors and References							
		1832–1841, Italian Seas [32]	1870, Adriatic Sea [33]	1879–1881, Sicilian Seas [34]	1883, Adriatic Sea [35]	1909, Central Thyrrenian Sea [36]	1956, Italian Seas [37]	1962, Italian Seas [38]	2012, Italian Seas [31]
Squalus acanthias	spiny dogfish	5	5	5	5	5	3	3	1
Squatina aculeata	sawback angel shark	na	na	na	na	na	1	2	1
Squatina oculata	smoothback angel shark	3	3	5	5	4	1	2	1
Squatina squatina	angel shark	5	5	3	5	na	1	2	1
Alopias vulpinus	common thresher shark	2	2	3	2	na	4	4	1
Mustelus asterias	starry smoothhound	5	5	5	5	na	5	5	2
Mustelus mustelus	common smoothhound	6	3	3	3	na	5	6	2
Mustelus punctulatus	blackspotted smoothhound	na	na	na	na	na	na	na	2
Prionace glauca	blue shark	na	3	na	2	4	4	5	3
Galeorhinus galeus	tope shark	5	2	5	5	3	5	3	1

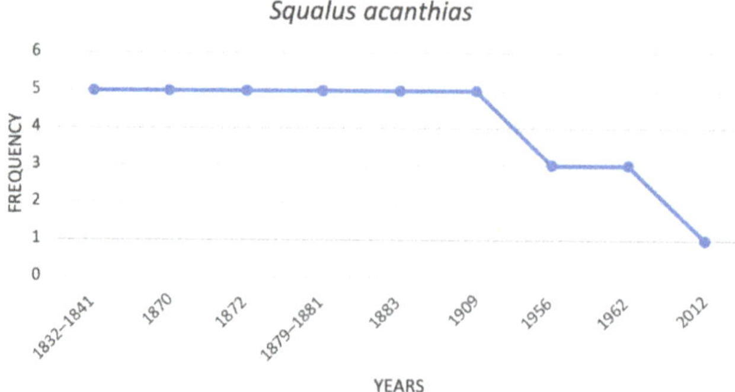

Figure 3. Temporal trend of *Squalus acanthias*.

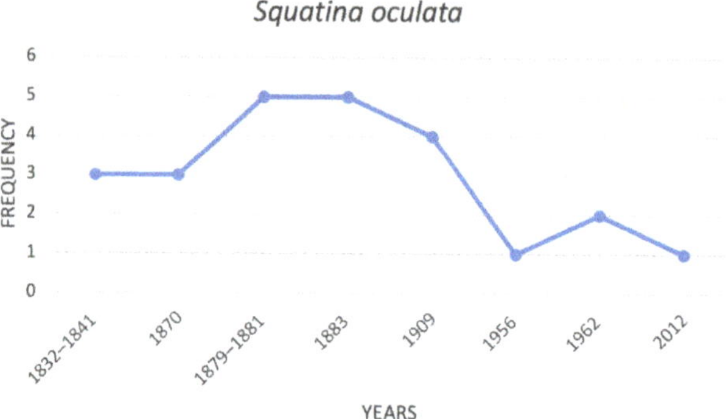

Figure 4. Temporal trend of *Squatina oculata*.

Squatina squatina was considered frequent along the Italian coast during the entire 19th century (Figure 5). Its populations seem to have collapsed during the first decades of the 20th century, and it became very rare from the 1940s onwards. *S. squatina* is currently listed as critically endangered (CR).

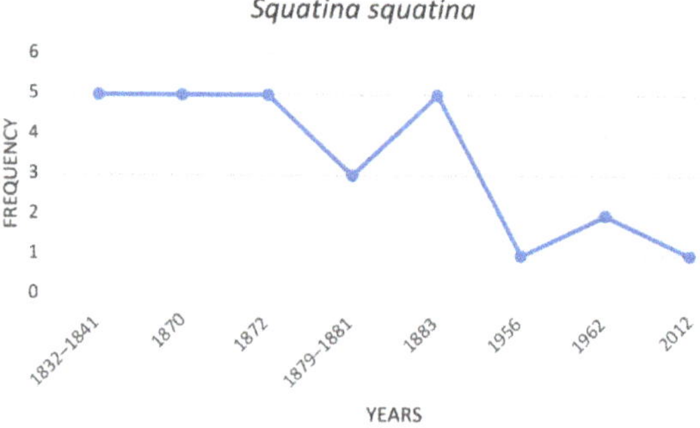

Figure 5. Temporal trend of *Squatina squatina*.

Alopias vulpinus was usually considered rare or uncommon in the Italian seas from the 19th century until the first decade after the end of WWII (Figure 6). In recent years *A. vulpinus* has been rarely recorded and is now listed as critically endangered (CR).

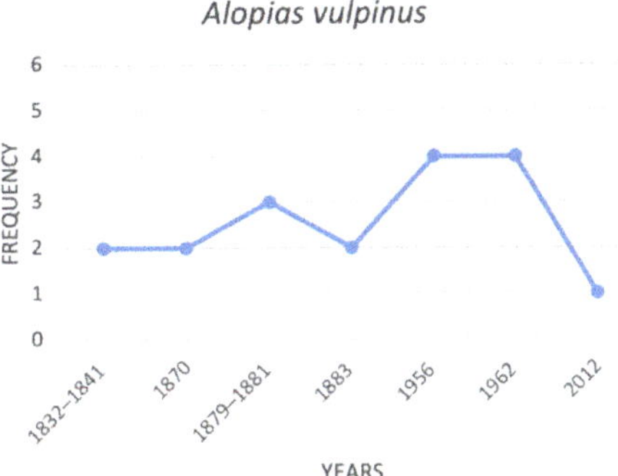

Figure 6. Temporal trend of *Alopias vulpinus*.

Mustelus asterias was considered very common in the Italian waters from the beginning of the 19th century until the 1950s (Figure 7). In the past 70 years, *M. asterias* has become rare and is now listed as endangered (EN).

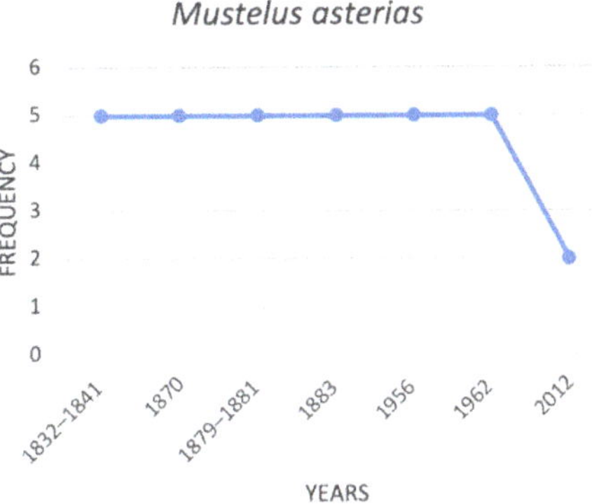

Figure 7. Temporal trend of *Mustelus asterias*.

Mustelus mustelus was considered very common at the beginning of the 19th century in the Italian waters, but its numbers decreased during the rest of the century (Figure 8). After WWII, *M. mustelus* was again considered very common. In the last 70 years it has been considered rare and is now listed as endangered (EN).

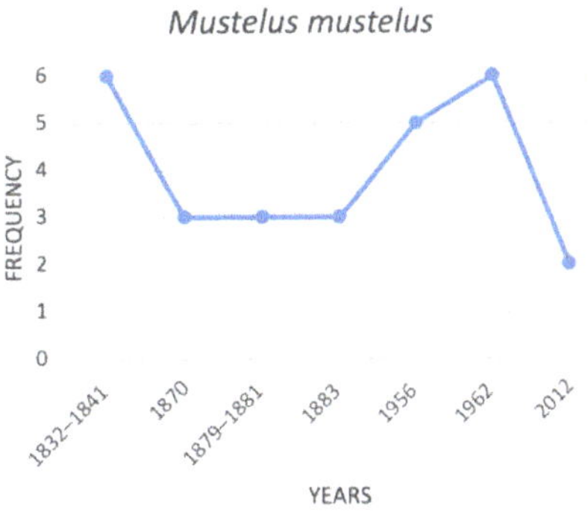

Figure 8. Temporal trend of *Mustelus mustelus*.

Prionace glauca was considered uncommon during the 19th century, although records indicated that it became more frequent from the beginning of the 20th century until the end of WWII (Figure 9). In the last 70 years this species has been considered rare and has been listed as vulnerable (VU).

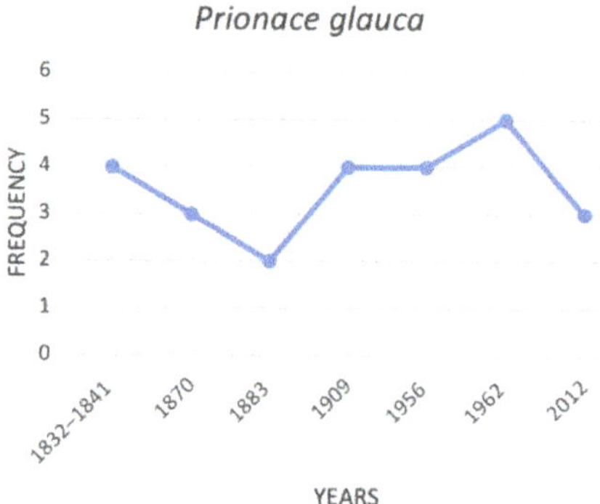

Figure 9. Temporal trend of *Prionace glauca*.

Galeorhinus galeus was considered common in Italian waters at the beginning of the 19th century. In 1870, records described it as rare. Thereafter, it was considered common again until the beginning of the 20th century. After WWII, *G. galeus* was again considered very common, but it has since been cited as less and less common and is currently classified as critically endangered (CR) in Italian waters (Figure 10).

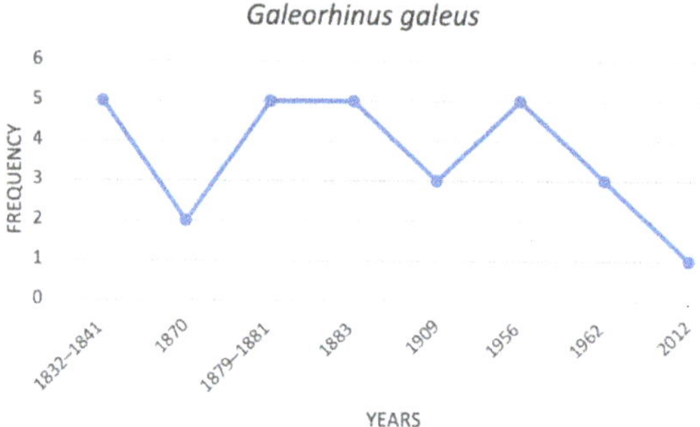

Figure 10. Temporal trend of *Galeorhinus galeus*.

4. Discussion

Even though applying historical ecology to marine environments has become more common in the last two decades [18,39,40], global studies focusing on cartilaginous fishes have been limited [41–46]. This lack of information is strongly evident for the Mediterranean Sea, [18,28–30,47–49]. Considering Italian waters, historical knowledge of cartilaginous fishes is not only limited, but is also fragmented. The available information does not apply to the entire peninsula, but rather, applies only for single local maritime districts. It therefore corresponds to observed variations in local abundances. Furthermore, research has been focused on a relatively recent temporal range, usually the last 70 years [27,50–52], and larger time windows are rarely considered [53], although sharks have been a common fisheries resource for a much longer period, as can be inferred from the vernacular names correctly attributed to individual species of sharks and the existence of specific fishing gear for such prey [54]. Starting from the IUCN Red List of the Italian cartilaginous fishes [31], the present study considered for the first time all the different threatened species of sharks along the Italian coasts, aiming to contribute to the construction of the first local baseline for these endangered vertebrates. Similarly to other marine megafauna, Chondrichthyans show a general worldwide decline [26] and, specifically, the selected species present significative negative trends, mainly in the last 70 years. Our results therefore highlight a reduction in abundance and species recorded.

4.1. Historical Trends of the Threatened Shark Species in the Italian Seas

Reviewing the scientific literature, the present study reveals that the spiny dogfish, *Squalus acanthias*, was considered very common in the Italian waters until the beginning of the 20th century, after which this species started to decrease rapidly. Ever since, *Squalus acanthias* have become extremely rare along the Italian peninsula, especially in the last 60 years. Similar situations were described in other maritime sectors, such as the northeastern Pacific and in the Atlantic Ocean [54–58]. This worldwide reduction has been caused by overexploitation [59], as well as habitat degradation and loss due to coastal development and pollution [60]. In some countries, the fishing of *S. acanthias* has been successfully managed, like in US waters [61], where local stocks were considered to be rebuilt in 2010 [62]. Similar initiatives may also be developed at the Italian level—in recent years, more studies focused on biological and ecological data related to this species have been carried out, providing preliminary information that is useful for its management [63–65]. Concerning the angel sharks of the genus *Squatina*, our research on the historical scientific literature confirmed their almost complete disappearance in Italy during the last century despite their regular occurrence until the early 1900s. This is in accordance with [66], who reported that angel sharks were common in the entire Mediterranean basin in the past, but are now detected only along the African and Middle East coasts, in the Aegean Sea and in Turkish waters. Indeed, these benthic cartilaginous fishes are actually confined to the less developed coastal areas of the Mediterranean. This is also the case in Italian waters, where the very rare modern records of angel sharks have been reported in the southern regions only [67–69], while in the northern regions these elasmobranchs have now disappeared [52]. Mediterranean angel sharks commonly inhabit the soft bottoms of very shallow coastal areas [70], where they are highly susceptible to trawlers and gillnets, as well as habitat degradation due to heavy human activity in coastal areas [52,66]. Currently, an international initiative to develop a Mediterranean action plan to protect and recover angel shark populations [71] has been started, based on recent and interesting citizen scientific research on these elasmobranchs [72]. These kinds of studies can be important in promoting international cooperation, which is fundamental for the conservation of threatened marine species like *Squatina* spp. Furthermore, they can play a fundamental role in expanding knowledge when combined with research into historical scientific bibliographic references like those reported in the present paper.

Alopias vulpinus, the thresher shark, is a typical pelagic shark, living in an offshore marine environment, and is rarely observed near the coastline [73]. In the Mediterranean, this species may be captured as by-catch in pelagic fisheries, mainly by longliners [74,75], but also during pair trawling fishing operations [64], as well as in offshore recreational fisheries [76]. In Italian waters, the uncommon detection of thresher sharks from the 19th century to the end of WWII can be related to the scarce chance of observing this high seas species in typical coastal fishing operations. Offshore pelagic fisheries require powerful engines to reach offshore fishing grounds, which were develop only during the past century [77]. In Italy, in particular, until the early 20th century, fishing was mainly practiced close to the coastline from boats powered by sails or oars, whereas motor fishing vessels were extremely rare [78,79]. On the contrary, after the forced stoppage due to WWII, the Italian fishing fleet underwent a technological change, mainly due to the development of powerful diesel engines, which caused a rapid increase and expansion of fishing efforts, which included pelagic fisheries [80]. The exploitation by the Italian fishery fleet of novel and pristine fishing grounds, like offshore pelagic environments, during the 1950s could be linked to the enhanced frequency of capture of *A. vulpinus* reported in the analyzed scientific literature. In the same way, the rapid decline in records of this species in the last 70 years may be connected to overexploitation of pelagic marine resources, including medium and small fishes, like mackerels, anchovies and sardines, which happen to be common preys of

A. vulpinus [81–83], but which are also one of the most important pelagic fishery resources in the Mediterranean Sea, where their capture numbers have decreased since the 1970s [49,83,84]. This may influence the upper trophic level, as was recently hypothesised for the thresher shark [85].

Our results highlight that the common smooth-hound, *Mustelus mustelus*, is very sensitive to fishing pressure in Italian waters. It nevertheless shows remarkable resilience, as also reported in other maritime sectors [86].There is a perception of a decrease in the frequency of observations of this species in the local historical scientific literature that corresponds to regular and continuous fishing pressure during the 19th century and the last 70 years. This observation is in line with a recent study on a century of fishery data on sharks of the genus *Mustelus* in the Mediterranean Sea, demonstrating the collapse of these marine vertebrates that occurred primarily in the last decades [28]. On the contrary, the recorded numbers of *M. mustelus* increased when the fishing effort partially or completely stopped during WWII and in the first years after the war. Similar trends were reported by D'Ancona for the Adriatic Sea [87,88] and have been recently confirmed for the entire Mediterranean basin by Colloca et al. [28] The dramatic collapse of *M. mustelus* and congeneric species in the Mediterranean basin and along the Italian coasts can be attributed mainly to the impact of direct fishing pressure [28], although it is also possible to relate it to the overfishing of its most common preys, which has also happened recently in the Mediterranean basin [89].

The blue shark, *Prionace glauca*, is one of the most wide-ranging shark species, found throughout all oceans [90,91]. It is considered to be one global population, with little or no differentiation within and between ocean basins [92–96]. Despite spending most of its life in pelagic environments, this species can also be found in coastal waters [70], a factor that might explain the occurrence of blue sharks during the 19th century and in the first decades of the 20th century in coastal Italian waters, when the Italian fishing effort was mainly concentrated close to the coastline [78,79]. With this exception, our study highlights that the historical trend of the blue shark is very similar to that of the thresher shark, *A. vulpinus*. Both species offer moderate resilience to fishing pressure [97,98]. The increase of recorded numbers of blue sharks during the 1950s can be related to the exploitation by the Italian fishery fleet of novel and pristine fishing grounds. As with the thresher shark, the rapid decline in the recorded numbers of *P. glauca* in the last decades may be connected to the over-exploitation of pelagic marine resources in the Mediterranean [99]. However, it can also be related to the degradation of the marine environments, as recent studies have demonstrated that blue sharks are strongly threatened by the presence of chemical pollution [100] and plastic debris [101].

The tope shark, *Galeorhinus galeus*, is a bento-pelagic species that shows a mainly coastal distribution, being found on the continental shelf and the upper slopes [70,102], while also being found offshore due to its extensive migrations [103]. Its coastal distribution allowed scientists to observe it at the beginning of the 19th century until the first decades of the 20th century, when Italian fishing effort was concentrated on the coastline [78,79]. During this period it was considered a rather common species [53,104] with some exceptions [33,36] which may be attributed to temporary and/or local low densities. Its gross decline started at the end of WWII, which was probably induced by the expansion of commercial fisheries and the advancement in fisheries technologies in Italian waters, resulting in higher fishing pressure on this species. Its decline in the past 70 years is in accordance with bottom trawl surveys from 1994 to 2009 [102,105], and other literature, in which it now appears to be locally extinct in the Tyrrhenian Sea [102,106] and is rarely captured in the Strait of Sicily [74,102].

4.2. Increasing Historical Knowledge to Improve Current Conservation

The heavy worldwide anthropization that occurred over the last two centuries has radically changed the structure and functioning of all ecosystems, including oceanic ecosystems [107]. Understanding historical ecological conditions is the first fundamental step in evaluating mitigation measures to best manage and conserve the marine environment [108]. Historical data provide essential information to better understand the recent human footprint on marine ecosystems and can be used not only to assess more accurately the baselines for different species, but also to develop management and conservation plans [26]. The present study provides new historical insights regarding the most threatened sharks according to the IUCN [31] present in Italian seas, a key area of one of the most anthropized maritime sectors [11]. The strong local decline of these predators in the last 200 years is evident, and can be attributed to different anthropic factors, such as bycatch, coastal habitat loss and environmental pollution. Considering all the cartilaginous fishes of Italian waters, it is worth noting that this scenario could be extended to many other species, as up to 52% of the Italian chondrichthyans are locally listed as data deficient (DD) by the IUCN, including many iconic species such as the great white shark and hammerhead sharks. Some studies have highlighted a general decline in the Mediterranean basin for some of the sharks included in this IUCN category [47,109], but the magnitude of reduction for most of the DD-designated species remains unknown, given the lack of quantitative evidence of historical abundance and distribution. Using historical data to understand how population abundances have changed over the years can further provide important information when assessing the vulnerability of sharks to climate change stressors. Ecological risk assessments (ERAs) are frequently used to estimate the influence of human actions on natural biological systems and processes [110,111], and can then be considered in integrated risk assessments for climate change. ERAs show particular effectiveness in the case of data-poor fishes, including elasmobranchs, which are often captured as bycatch in fisheries [112,113]. Understanding if a species has been considered rare over a long time or has only recently been considered rare due to overexploitation and anthropic effects is an interesting factor to consider when assessing the risk of different changes upon species. This is because rare chondrichthyans may have reduced phenotypic variations and are therefore more susceptible to changes [113]. From this point of view, improved monitoring of sightings and statistical analysis of the results of threatened and DD sharks in Italian seas—along with enhanced efforts in searching and analyzing historical data—are essential in order to increase our understanding of these important and vulnerable marine organisms [46]. At the same time, a precautionary approach should be taken by marine stakeholders to reduce the current anthropic impact on chondrichthyans. This can be achieved through the use and distribution of bycatch reduction devices, the implementation of good fishing practices, the release of still-living accidentally caught specimens and eliminating the production of new marine litter.

Author Contributions: Conceptualization, F.L.L., E.S. and M.B.; methodology, F.L.L., E.S., A.T., P.N.P. and M.B.; investigation, F.L.L., E.S., A.T., A.R.M., M.S., P.N.P. and M.B.; data curation, F.L.L., E.S., T.M.D. and M.B.; writing-original, draft preparation, F.L.L., E.S., T.M.D. and M.B.; writing—review and editing, F.L.L., E.S., M.S. and M.B. All authors have read and agreed to the published version of the manuscript.

Funding: F.L.L. was supported by a Ph.D. grant co-funded by the Stazione Zoologica Anton Dohrn and the University of Calabria. T.M.D was supported by a research fellowship funded the Stazione Zoologica Anton Dohrn in the framework of the LIFE ELIFE Project—Elasmobranchs Low-Impact Fishing Experience (LIFE18 NAT/IT/000846).

Acknowledgments: This research was partially supported and carried out in the framework of the LIFE ELIFE Project—Elasmobranchs Low-Impact Fishing Experience (LIFE18 NAT/IT/000846), funded by the contribution of the LIFE financial instrument of the European Community. It does not necessarily reflect the European Commission's views and in no way anticipates future policy. This support is greatly acknowledged This study was also part of the Ph.D. project of F.L.L. The authors are grateful to Gianni Giglio (DiBEST, University of Calabria) for technical collaboration during data analysis.

Conflicts of Interest: The authors declare no conflict of interest.

References

1. Dulvy, N.K.; Fowler, S.L.; A Musick, J.; Cavanagh, R.D.; Kyne, P.M.; Harrison, L.R.; Carlson, J.K.; Davidson, L.N.; Fordham, S.V.; Francis, M.P.; et al. Extinction risk and conservation of the world's sharks and rays. *eLife* **2014**, *3*, e00590. [CrossRef]
2. Simpfendorfer, C.A.; Kyne, P.M. Limited potential to recover from overfishing raises concerns for deep-sea sharks, rays and chimaeras. *Environ. Conserv.* **2009**, *36*, 97–103. [CrossRef]
3. Queiroz, M.M.; Pereira, S.C.F. Intention to adopt big data in supply chain management: A Brazilian perspective. *Rev. Adm. Empresas* **2019**, *59*, 389–401. [CrossRef]

4. Simpfendorfer, C.A.; Dulvy, N.K. Bright spots of sustainable shark fishing. *Curr. Biol.* **2017**, *27*, R97–R98. [CrossRef]

5. Cortés, E.; Domingo, A.; Miller, P.; Forselledo, R.; Mas, F.; Arocha, F.; Campana, S.E.; Coelho, R.; da Silva, C.; Hazin, F.; et al. Expanded ecological risk assessment of pelagic sharks caught in Atlantic pelagic longline fisheries. *Collect. Vol. Sci. Pap. ICCAT* **2015**, *71*, 2637–2688.

6. Dulvy, N.K.; Simpfendorfer, C.A.; Davidson, L.N.; Fordham, S.V.; Bräutigam, A.; Sant, G.; Welch, D.J. Challenges and Priorities in Shark and Ray Conservation. *Curr. Biol.* **2017**, *27*, R565–R572. [CrossRef] [PubMed]

7. Pauly, D. Anecdotes and the shifting baseline syndrome of fisheries. *Trends Ecol. Evol.* **1995**, *10*, 430. [CrossRef]

8. Sáenz-Arroyo, A.; Roberts, C.M.; Torre, J.; Cariño-Olvera, M. Using fishers' anecdotes, naturalists' observations and grey literature to reassess marine species at risk: The case of the Gulf grouper in the Gulf of California, Mexico. *Fish Fish.* **2005**, *6*, 121–133. [CrossRef]

9. Osio, G.C.; Orio, A.; Millar, C.P. Assessing the vulnerability of Mediterranean demersal stocks and predicting exploitation status of un-assessed stocks. *Fish. Res.* **2015**, *171*, 110–121. [CrossRef]

10. Dulvy, N.K.; Allen, D.J.; Ralph, G.M.; Walls, R.H.L. *The Conservation Status of Sharks, Rays and Chimaeras in the Mediterranean Sea [Brochure]*; IUCN Centre for Mediterranean Cooperation: Malaga, Spain, 2016; 14p.

11. Coll, M.; Piroddi, C.; Steenbeek, J.; Kaschner, K.; Lasram, F.B.R.; Aguzzi, J.; Ballesteros, E.; Bianchi, C.N.; Corbera, J.; Dailianis, T.; et al. The biodiversity of the Mediterranean Sea: Estimates, patterns, and threats. *PLoS ONE* **2010**, *5*, e11842. [CrossRef] [PubMed]

12. Katsanevakis, S.; Levin, N.; Coll, M.; Giakoumi, S.; Shkedi, D.; Mackelworth, P.; Levy, R.; Velegrakis, A.; Koutsoubas, D.; Carić, H.; et al. Marine conservation challenges in an era of economic crisis and geopolitical instability: The case of the Mediterranean Sea. *Mar. Policy* **2015**, *51*, 31–39. [CrossRef]

13. Bianchi, C.N.; Morri, C. Global sea warming and "tropicalization" of the Mediterranean Sea: Biogeographic and ecological aspects. *Biogeogr. J. Integr. Biogeogr.* **2003**, *24*. [CrossRef]

14. Bianchi, C.N.; Morri, C. Marine biodiversity of the Mediterranean Sea: Situation, problems and prospects for future research. *Mar. Pollut. Bull.* **2000**, *40*, 367–376. [CrossRef]

15. Guidetti, P.; Milazzo, M.; Bussotti, S.; Molinari, A.; Murenu, M.; Pais, A.; Spanò, N.; Balzano, R.; Agardy, T.; Boero, F.; et al. Italian marine reserve effectiveness: Does enforcement matter? *Biol. Conserv.* **2008**, *141*, 699–709. [CrossRef]

16. Szabó, P. Historical ecology: Past, present and future. *Biol. Rev.* **2015**, *90*, 997–1014. [CrossRef]

17. Balée, W. The Research Program of Historical Ecology. *Annu. Rev. Anthropol.* **2006**, *35*, 75–98. [CrossRef]

18. McClenachan, L.; Cooper, A.B.; McKenzie, M.G.; Drew, J.A. The importance of surprising results and best practices in historical ecology. *BioScience* **2015**, *65*, 932–939. [CrossRef]

19. Lotze, H.K.; Hoffmann, R.; Erlandson, J. Lessons from historical ecology and management. *Mar. Ecosyst. Based Manag. Sea Ideas Observ. Prog. Study Seas* **2014**, *16*, 17–55.

20. Engelhard, G.H.; Lynam, C.P.; Garcia-Carreras, B.; Dolder, P.J.; Mackinson, S. Effort reduction and the large fish indicator: Spatial trends reveal positive impacts of recent European fleet reduction schemes. *Environ. Conserv.* **2015**, *42*, 227–236. [CrossRef]

21. Sanchez, G.M.; Gobalet, K.W.; Jewett, R.; Cuthrell, R.Q.; Grone, M.; Engel, P.M.; Lightfoot, K.G. The historical ecology of central California coast fishing: Perspectives from Point Reyes National Seashore. *J. Archaeol. Sci.* **2018**, *100*, 1–15. [CrossRef]

22. Trindade-Santos, I.; Moyes, F.; Magurran, A.E. Global change in the functional diversity of marine fisheries exploitation over the past 65 years. *Proc. R. Soc. B* **2020**, *287*, 20200889. [CrossRef] [PubMed]

23. Guidetti, P.; Micheli, F. Ancient art serving marine conservation. *Front. Ecol. Environ.* **2011**, *9*, 374–375. [CrossRef]

24. Lotze, H.K.; Coll, M.; Magera, A.M.; Ward-Paige, C.; Airoldi, L. Recovery of marine animal populations and ecosystems. *Trends Ecol. Evol.* **2011**, *26*, 595–605. [CrossRef]

25. Fortibuoni, T.; Libralato, S.; Raicevich, S.; Giovanardi, O.; Solidoro, C. Coding early naturalists' accounts into long-term fish community changes in the Adriatic Sea (1800–2000). *PLoS ONE* **2010**, *5*, e15502. [CrossRef] [PubMed]

26. McClenachan, L.; Ferretti, F.; Baum, J.K. From archives to conservation: Why historical data are needed to set baselines for marine animals and ecosystems. *Conserv. Lett.* **2012**, *5*, 349–359. [CrossRef]

27. Ferretti, F.; Crowder, L.B.; Micheli, F. Using Disparate Datasets to Reconstruct Historical Baselines of Animal Populations. In *Marine Historical Ecology in Conservation: Applying the Past to Manage for the Future*; University of California Press: Oakland, CA, USA, 2014; pp. 63–85.
28. Colloca, F.; Enea, M.; Ragonese, S.; Di Lorenzo, M. A century of fishery data documenting the collapse of smooth-hounds (*Mustelus* spp.) in the Mediterranean Sea. *Aquat. Conserv. Mar. Freshw. Ecosyst.* **2017**, *27*, 1145–1155. [CrossRef]
29. Mojetta, A.R.; Travaglini, A.; Scacco, U.; Bottaro, M. Where sharks met humans: The Mediterranean Sea, history and myth of an ancient interaction between two dominant predators. *Reg. Stud. Mar. Sci.* **2018**, *21*, 30–38. [CrossRef]
30. Bargnesi, F.; Gridelli, S.; Cerrano, C.; Ferretti, F. Reconstructing the history of the sand tiger shark (*Carcharias taurus*) in the Mediterranean Sea. *Aquat. Conserv. Mar. Freshw. Ecosyst.* **2020**, *30*, 915–927. [CrossRef]
31. Rondinini, C.; Battistoni, A.; Peronace, V.; Teofili, C. *Lista Rossa IUCN dei Vertebrati Italiani*; Comitato Italiano IUCN e Ministero dell'Ambiente e della Tutela del Territorio e del Mare: Roma, Italy, 2012; Volume 56.
32. Bonaparte, C.L. Iconografia della fauna italica per le quattro classi degli animali vertebrati. Tomo III. Pesci. *Roma Fasc* **1832**, *1*, 1–6.
33. Ninni, A.P. *Enumerazione dei Pesci delle Lagune e Golfo di Venezia con Note per il Dott. Alessandro Ninni*; Soliani: Modena, Italy, 1870.
34. Doderlein, P. *Manuale Ittiologico del Mediterraneo: Ossia Sinossi Metodica delle Varie Specie di Pesci Riscontrate nel Mediterraneo ed in Particolare nei Mari di Sicilia*; Tip. del Giornale di Sicilia: Palermo, Italy, 1879.
35. Faber, G.L. *The Fisheries of the Adriatic and the Fish Thereof*; Bernard Quaritch: London, UK, 1883.
36. Lo Bianco, S. Biological notice with special reference to the period of sexual maturity of the animals of the Bay of Naples. *Mitth. Zool. Stat. Neapel* **1909**, *19*, 513–761.
37. Tortonese, E. *Fauna d'Italia: Vol. II—Leptocardia Ciclostomata Selachi*; Editoriale Calderini: Bologna, Italy, 1956.
38. Bini, G. *Atlante Dei Pesci Delle Coste Italiane*; Mondo Sommerso; Editoriale Olimpia: Firenze, Italy, 1962.
39. Jackson, J.B.C.; Alexander, K.E.; Sala, E. *Shifting Baselines: The Past and the Future of Ocean Fisheries*; Island Press: Washington, DC, USA, 2011; p. 312.
40. Caswell, B.; Klein, E.S.; Alleway, H.K.; Ball, J.E.; Botero, J.; Cardinale, M.; Eero, M.; Engelhard, G.H.; Fortibuoni, T.; Giraldo, A.-J.; et al. Something old, something new: Historical perspectives provide lessons for blue growth agendas. *Fish Fish.* **2020**, *21*, 774–796. [CrossRef]
41. Baum, J.K.; Myers, R.A.; Kehler, D.G.; Worm, B.; Harley, S.J.; Doherty, P.A. Collapse and Conservation of Shark Populations in the Northwest Atlantic. *Science* **2003**, *299*, 389–392. [CrossRef] [PubMed]
42. Myers, R.A.; Worm, B. Rapid worldwide depletion of predatory fish communities. *Nature* **2003**, *423*, 280–283. [CrossRef]
43. Myers, R.A.; Worm, B. Extinction, survival or recovery of large predatory fishes. *Philos. Trans. R. Soc. B Biol. Sci.* **2005**, *360*, 13–20. [CrossRef] [PubMed]
44. Ferretti, F.; Worm, B.; Britten, G.L.; Heithaus, M.R.; Lotze, H.K. Patterns and ecosystem consequences of shark declines in the ocean. *Ecol. Lett.* **2010**, *13*, 1055–1071. [CrossRef] [PubMed]
45. Roff, G.; Brown, C.J.; Priest, M.A.; Mumby, P.J. Decline of coastal apex shark populations over the past half century. *Commun. Boil.* **2018**, *1*, 223. [CrossRef] [PubMed]
46. Martínez-Candelas, I.A.; Pérez-Jiménez, J.C.; Espinoza-Tenorio, A.; McClenachan, L.; Méndez-Loeza, I. Use of historical data to assess changes in the vulnerability of sharks. *Fish. Res.* **2020**, *226*, 105526. [CrossRef]
47. Ferretti, F.; Myers, R.A.; Serena, F.; Lotze, H.K. Loss of large predatory sharks from the Mediterranean Sea. *Conserv. Biol.* **2008**, *22*, 952–964. [CrossRef] [PubMed]
48. Ferretti, F.; Morey Verd, G.; Seret, B.; Sulić Šprem, J.; Micheli, F. Falling through the cracks: The fading history of a large iconic predator. *Fish Fish.* **2016**, *17*, 875–889. [CrossRef]
49. Fortibuoni, T.; Libralato, S.; Arneri, E.; Giovanardi, O.; Solidoro, C.; Raicevich, S. Fish and fishery historical data since the 19th century in the Adriatic Sea, Mediterranean. *Sci. Data* **2017**, *4*, 170104. [CrossRef]
50. Barausse, A.; Correale, V.; Curkovic, A.; Finotto, L.; Riginella, E.; Visentin, E.; Mazzoldi, C. The role of fisheries and the environment in driving the decline of elasmobranchs in the northern Adriatic Sea. *ICES J. Mar. Sci.* **2014**, *71*, 1593–1603. [CrossRef]
51. Mazzoldi, C.; Sambo, A.; Riginella, E. The Clodia database: A long time series of fishery data from the Adriatic Sea. *Sci. Data* **2014**, *1*, 140018. [CrossRef] [PubMed]

52. Fortibuoni, T.; Borme, D.; Franceschini, G.; Giovanardi, O.; Raicevich, S. Common, rare or extirpated? Shifting baselines for common angelshark, *Squatina squatina* (Elasmobranchii: Squatinidae), in the Northern Adriatic Sea (Mediterranean Sea). *Hydrobiologia* **2016**, *772*, 247–259. [CrossRef]

53. Psomadakis, P.N.; Maio, N.; Vacchi, M. The chondrichthyan biodiversity in the Gulf of Naples (SW Italy, Tyrrhenian Sea): An historical overview. *Cybium* **2009**, *33*, 199–209.

54. Targioni-Tozzetti, A. *La Pesca in Italia (Vol. 2)*; R. Istituto Sordo-Muti: Genoa, Italy, 1874.

55. Alverson, D.L.; Stansby, M.E. *The Spiny Dogfish (Squalus acanthias) in the Northeastern Pacific (No. 447)*; US Department of the Interior, Fish and Wildlife Service, Bureau of Commercial Fisheries: Washington, DC, USA, 1963.

56. Alonso, M.K.; Crespo, E.A.; García, N.A.; Pedraza, S.N.; Mariotti, P.A.; Mora, N.J. Fishery and ontogenetic driven changes in the diet of the spiny dogfish, Squalus acanthias, in Patagonian waters, Argentina. *Environ. Biol. Fish.* **2002**, *63*, 193–202. [CrossRef]

57. King, J.R.; McFarlane, G.A. Trends in Abundance of Spiny Dogfish in the Strait of Georgia, 1980–2005. In *Biology and Management of Dogfish Sharks*; American Fisheries Society: Bethesda, MD, USA, 2009; pp. 89–100.

58. Belleggia, M.; Figueroa, D.E.; Sánchez, F.; Bremec, C. Long-term changes in the spiny dogfish (*Squalus acanthias*) trophic role in the southwestern Atlantic. *Hydrobiologia* **2012**, *684*, 57–67. [CrossRef]

59. Fordham, S.; Dolan, C. A case study in international shark conservation: The convention on international trade in endangered species and the spiny dogfish. *Gold. Gate UL Rev.* **2004**, *34*, 531.

60. Fowler, S.; Raymakers, C.; Grimm, U. *Trade in and Conservation of Two Shark Species, Porbeagle (Lamna nasus) and Spiny Dogfish (Squalus acanthias)*; Bundesamt für Naturschutz (BfN): Bonn, Germany, 2004.

61. MAFMC. Mid-Atlantic Fishery Management Council. In *Spiny Dogfish Fishery Management Plan*; MAFMC: Dover, DE, USA, 1999.

62. Dell'Apa, A.; Bangley, C.W.; Rulifson, R.A. Who let the dogfish out? A review of management and socio-economic aspects of spiny dogfish fisheries. *Rev. Fish Biol. Fish.* **2015**, *25*, 273–295. [CrossRef]

63. Bellodi, A.; Porcu, C.; Cau, A.; Marongiu, M.F.; Melis, R.; Mulas, A.; Pesci, P.; Follesa, M.C.; Cannas, R. Investigation on the genus Squalus in the Sardinian waters (Central-Western Mediterranean) with implications on its management. *Mediterr. Mar. Sci.* **2018**, *19*, 256–272. [CrossRef]

64. Bonanomi, S.; Pulcinella, J.; Fortuna, C.M.; Moro, F.; Sala, A. Elasmobranch bycatch in the Italian Adriatic pelagic trawl fishery. *PLoS ONE* **2018**, *13*, e0191647. [CrossRef]

65. Bargione, G.; Donato, F.; La Mesa, M.; Mazzoldi, C.; Riginella, E.; Vasapollo, C.; Virgili, M.; Lucchetti, A. Life-history traits of the spiny dogfish Squalus acanthias in the Adriatic Sea. *Sci. Rep.* **2019**, *9*, 14317. [CrossRef] [PubMed]

66. Lawson, J.M.; Pollom, R.A.; Gordon, C.A.; Barker, J.; Meyers, E.K.; Zidowitz, H.; Ellis, J.R.; Bartolí, Á.; Morey, G.; Fowler, S.L.; et al. Extinction risk and conservation of critically endangered angel sharks in the Eastern Atlantic and Mediterranean Sea. *ICES J. Mar. Sci.* **2020**, *77*, 12–29. [CrossRef]

67. Ragonese, S.; Vitale, S.; Dimech, M.; Mazzola, S. Abundances of demersal sharks and chimaera from 1994–2009 scientific surveys in the central Mediterranean Sea. *PLoS ONE* **2013**, *8*, e74865. [CrossRef] [PubMed]

68. Cavallaro, M.; Ammendolia, G.; Navarra, E. Finding of a rare *Squatina squatina* (Linnaeus, 1758) (Chondrichthyes: Squatinidae) along the Tyrrhenian coast of the Strait of Messina and its maintenance in an aquarium. *Mar. Biodivers. Rec.* **2015**, *8*. [CrossRef]

69. Zava, B.; Fiorentino, F.; Serena, F. Occurrence of juveniles *Squatina oculata* Bonaparte, 1840 (Elasmobranchii: Squatinidae) in the Strait of Sicily (Central Mediterranean). *Cybium Int. J. Ichthyol.* **2016**, *1840*, 341–343.

70. Serena, F. *Field Identification Guide to the Sharks and Rays of the Mediterranean and Black Sea*; Food and Agriculture Organization of the United Nations: Rome, Italy, 2005.

71. Gordon, C.A.; Hood, A.R.; Al Mabruk, S.A.A.; Barker, J.; Bartolí, A.; Ben Abdelhamid, S.; Bradai, M.N.; Dulvy, N.K.; Fortibuoni, T.; Giovos, I.; et al. *Mediterranean Angel Sharks: Regional Action Plan*; The Shark Trust: Plymouth, UK, 2019.

72. Giovos, I.; Stoilas, V.; Al-Mabruk, S.A.; Doumpas, N.; Marakis, P.; Maximiadi, M.; Moutopoulos, D.K.; Kleitou, P.; Keramidas, I.; Tiralongo, F.; et al. Integrating local ecological knowledge, citizen science and long-term historical data for endangered species conservation: Additional records of angel sharks (Chondrichthyes: Squatinidae) in the Mediterranean Sea. *Aquat. Conserv. Mar. Freshw. Ecosyst.* **2019**, *29*, 881–890. [CrossRef]

73. Smith, S.E.; Rasmussen, R.C.; Ramon, D.A.; Cailliet, G.M. The Biology and Ecology of Thresher Sharks (Alopiidae). In *Sharks of the Open Ocean: Biology, Fisheries and Conservation*; Wiley: Hoboken, NJ, USA, 2008; pp. 60–68.

74. Megalofonou, P.; Yannopoulos, C.; Damalas, D.; De Metrio, G.; De Florio, M. Incidental catch and estimated discards of pelagic sharks from the swordfish and tuna fisheries in the Mediterranean Sea. *Fish. Bull.* **2005**, *103*, 620–634.

75. Garibaldi, F. By-catch in the mesopelagic swordfish longline fishery in the Ligurian Sea (Western Mediterranean). *Collect. Vol. Sci. Pap. ICCAT* **2015**, *71*, 1495–1498.

76. Panayiotou, N.; Porsmoguer, S.B.; Moutopoulos, D.K.; Lloret, J. Offshore recreational fisheries of large vulnerable sharks and teleost fish in the Mediterranean Sea: First information on the species caught. *Mediterr. Mar. Sci.* **2020**, *21*, 222–227. [CrossRef]

77. Watson, J.W.; Kerstetter, D.W. Pelagic longline fishing gear: A brief history and review of research efforts to improve selectivity. *Mar. Technol. Soc. J.* **2006**, *40*, 6–11. [CrossRef]

78. Sassu, N.; Cannas, A.; Ferretti, M. *Gli Attrezzi da Pesca in Uso nelle Marinerie Italiane*; UNIMAR: Rome, Italy, 2001; Volume 16, p. 81.

79. Ferretti, M. *Classificazione e Descrizione Degli Attrezzi da Pesca in Uso Nelle Marinerie Italiane con Particolare Referimento al Loro Impatto Ambientale*; ICRAM: Rome, Italy, 2002.

80. Cataudella, S.; Spagnolo, M. *Lo Stato della Pesca e Dell'Acquacoltura nei Mari Italiani*; Ministero delle Politiche Agricole Alimentari e Forestali: Rome, Italy, 2011; 877p.

81. Preti, A.; Smith, S.E.; Ramon, D.A. Diet differences in the thresher shark (Alopias vulpinus) during transition from a warm-water regime to a cool-water regime off California-Oregon, 1998–2000. *Calif. Cooper. Ocean. Fish. Investig. Rep.* **2004**, *45*, 118.

82. Preti, A.; Soykan, C.U.; Dewar, H.; Wells, R.D.; Spear, N.; Kohin, S. Comparative feeding ecology of shortfin mako, blue and thresher sharks in the California Current. *Environ. Biol. Fish.* **2012**, *95*, 127–146. [CrossRef]

83. Balestra, V.; Boero, F.; Carli, A. Andamento del pescato della tonnarella di Camogli dal 1950 al 1974. Valutazioni bio-statistiche. *Boll. Pesca Piscic. Idrobiol.* **1976**, *31*, 2.

84. Cattaneo-Vietti, R. Structural changes in Mediterranean marine communities: Lessons from the Ligurian Sea. *Rend. Lincei. Sci. Fis. Nat.* **2018**, *29*, 515–524.

85. Finotto, L.; Barausse, A.; Mazzoldi, C. In search of prey: The occurrence of Alopias vulpinus (Bonnaterre, 1788) in the northern Adriatic Sea and its interactions with fishery. *Acta Adriat.* **2016**, *57*, 295–304.

86. Sguotti, C.; Lynam, C.P.; García-Carreras, B.; Ellis, J.R.; Engelhard, G.H. Distribution of skates and sharks in the North Sea: 112 years of change. *Glob. Chang. Boil.* **2016**, *22*, 2729–2743. [CrossRef]

87. D'Ancona, U. Dell'Influenza della Stasi Peschereccia del Periodo 1914–1918 sul Patrimonio Ittico Dell'Alto Adriatico. In *Regio Comitato Talassografico Italiano, Memoria CXXVI*; Officine Grafiche Carlo Ferrari: Venice, Italy, 1926; pp. 1–95.

88. D'Ancona, U. Rilievi statistici sulla pesca nell'alto Adriatico. *Istit. Ven. Sci. Lett. Arti* **1949**, *108*, 41–53.

89. Di Lorenzo, M.; Vizzini, S.; Signa, G.; Andolina, C.; Palo, G.B.; Gristina, M.; Mazzoldi, C.; Colloca, F. Ontogenetic trophic segregation between two threatened smooth-hound sharks in the Central Mediterranean Sea. *Sci. Rep.* **2020**, *10*, 11011. [CrossRef]

90. Last, P.R.; Stevens, J.D. *Sharks and Rays of Australia*, 2nd ed.; CSIRO Publishing: Melbourne, Australia, 2009.

91. Ebert, D.A.; Fowler, S.L.; Compagno, L.J. *Sharks of the World: A Fully Illustrated Guide*; Wild Nature Press: Princeton, NJ, USA, 2013.

92. Ovenden, J.R.; Kashiwagi, T.; Broderick, D.; Giles, J.; Salini, J. The extent of population genetic subdivision differs among four co-distributed shark species in the Indo-Australian archipelago. *BMC Evol. Boil.* **2009**, *9*, 40. [CrossRef]

93. King, J.R.; Wetklo, M.; Supernault, J.; Taguchi, M.; Yokawa, K.; Sosa-Nishizaki, O.; Withler, R.E. Genetic analysis of stock structure of blue shark (*Prionace glauca*) in the north Pacific ocean. *Fish. Res.* **2015**, *172*, 181–189. [CrossRef]

94. Leone, A.; Urso, I.; Damalas, D.; Martinsohn, J.; Zanzi, A.; Mariani, S.; Sperone, E.; Micarelli, P.; Garibaldi, F.; Megalofonou, P.; et al. Genetic differentiation and phylogeography of Mediterranean-North Eastern Atlantic blue shark (*Prionace glauca*, L. 1758) using mitochondrial DNA: Panmixia or complex stock structure? *PeerJ* **2017**, *5*, e4112. [CrossRef] [PubMed]

95. Veríssimo, A.; Sampaio, Í.; McDowell, J.R.; Alexandrino, P.; Mucientes, G.; Queiroz, N.; Da Silva, C.; Jones, C.S.; Noble, L.R. World without borders—Genetic population structure of a highly migratory marine predator, the blue shark (*Prionace glauca*). *Ecol. Evol.* **2017**, *7*, 4768–4781.

96. Bailleul, D.; MacKenzie, A.; Sacchi, O.; Poisson, F.; Bierne, N.; Arnaud-Haond, S. Large-scale genetic panmixia in the blue shark (*Prionace glauca*): A single worldwide population, or a genetic lag-time effect of the "grey zone" of differentiation? *Evol. Appl.* **2018**, *11*, 614–630. [CrossRef]

97. Campana, S.E.; Marks, L.; Joyce, W.; Kohler, N.E. Effects of recreational and commercial fishing on blue sharks (*Prionace glauca*) in Atlantic Canada, with inferences on the North Atlantic population. *Can. J. Fish. Aquat. Sci.* **2006**, *63*, 670–682. [CrossRef]

98. Froese, R.; Garilao, C.; Winker, H.; Coro, G.; Demirel, N.; Tsikliras, A.; Dimarchopoulou, D.; Scarcella, G.; Sampang-Reyes, A. *Exploitation and Status of European Stocks*; Oceana: Washington, DC, USA, 2016.

99. Biton-PorSmoguer, S.; LLoret, J. Potentially unsustainable fisheries of a critically-endangered pelagic shark species: The case of the Blue shark (*Prionace glauca*) in the Western Mediterranean Sea. *Cybium* **2018**, *42*, 299–302.

100. Storelli, M.M.; Barone, G.; Storelli, A.; Marcotrigiano, G.O. Levels and congener profiles of PCBs and PCDD/Fs in blue shark (*Prionace glauca*) liver from the South-Eastern Mediterranean Sea (Italy). *Chemosphere* **2011**, *82*, 37–42. [CrossRef] [PubMed]

101. Bernardini, I.; Garibaldi, F.; Canesi, L.; Fossi, M.C.; Baini, M. First data on plastic ingestion by blue sharks (*Prionace glauca*) from the Ligurian Sea (North-Western Mediterranean Sea). *Mar. Pollut. Bull.* **2018**, *135*, 303–310. [CrossRef]

102. Colloca, F.; Scannela, D.; Geraci, M.L.; Falsone, F.; Giusto, B.; Vitale, S.; Di Lorenzo, M.; Bono, G. British sharks in Sicily: Records of long-distance migration of tope shark (*Galeorhinus galeus*) from the north-eastern Atlantic to the Mediterranean Sea. *Mediterr. Mar. Sci.* **2019**, *20*, 309–313. [CrossRef]

103. United Nations Environment (UNEP). Proposal for the Inclusion of the Tope Shark (*Galeorhinus galeus*) in Appendix II of the Convention. In Proceedings of the 4th Meeting of the Sessional Committee of the CMS Scientific Council (ScC-SC4), Bonn, Germany, 12–15 November 2019. UNEP/CMS/COP13/Doc. 27.1.10.

104. D'ancona, U.; Razzauti, A. Pesci e Pesca nelle acque dell'Arcipelago Toscano. *Comm. Int. Exp. Sci. Médit. Rapp. Pro. Verb.* **1937**, *11*, 129.

105. Relini, G.; Mannini, A.; De Ranieri, S.; Bitetto, I.; Follesa, M.C.; Gancitano, V.; Manfredi, C.; Casciaro, L.; Sion, L. Chondrichthyes caught during the medits surveys in Italian waters. *Biol. Mar. Mediterr.* **2010**, *17*, 186–204.

106. Ferretti, F.; Myers, R.A.; Sartor, P.; Serena, F. Long term dynamics of the chondrichthyan fish community in the upper Tyrrhenian Sea. *ICES CM* **2005**, *25*, 1–34.

107. Ferretti, F.; Curnick, D.; Liu, K.; Romanov, E.V.; Block, B.A. Shark baselines and the conservation role of remote coral reef ecosystems. *Sci. Adv.* **2018**, *4*, eaaq0333. [CrossRef] [PubMed]

108. Drew, J.; Philipp, C.; Westneat, M.W. Shark Tooth Weapons from the 19th Century Reflect Shifting Baselines in Central Pacific Predator Assemblies. *PLoS ONE* **2013**, *8*, e59855. [CrossRef] [PubMed]

109. Moro, S.; Jona-Lasinio, G.; Block, B.; Micheli, F.; De Leo, G.; Serena, F.; Bottaro, M.; Scacco, U.; Ferretti, F. Abundance and distribution of the white shark in the Mediterranean Sea. *Fish Fish.* **2020**, *21*, 338–349. [CrossRef]

110. Burgman, M. *Risk and Decisions for Conservation and Environmental Management*; Cambridge University Press: Cambridge, UK, 2005.

111. Gallagher, A.J.; Kyne, P.M.; Hammerschlag, N. Ecological risk assessment and its application to elasmobranch conservation and management. *J. Fish Biol.* **2012**, *80*, 1727–1748. [CrossRef]

112. Milton, D.A. Assessing the susceptibility to fishing of populations of rare trawl bycatch: Sea snakes caught by Australia's Northern Prawn Fishery. *Biol. Conserv.* **2001**, *101*, 281–290. [CrossRef]

113. Chin, A.; Kyne, P.M.; Walkers, T.I.; McAucley, R.B. An integrated risk assessment for climate change: Analyzing the vulnerability of sharks and rays on Australia's Great Barrier Reef. *Glob. Chang. Biol.* **2010**, *16*, 1936–1953. [CrossRef]

Communication

A Mediterranean Monk Seal Pup on the Apulian Coast (Southern Italy): Sign of an Ongoing Recolonisation?

Tatiana Fioravanti [1], Andrea Splendiani [1], Tommaso Righi [1], Nicola Maio [2],
Sabrina Lo Brutto [3], Antonio Petrella [4] and Vincenzo Caputo Barucchi [1,*]

1 Dipartimento di Scienze della Vita e dell'Ambiente (DiSVA), Università Politecnica delle Marche, Via Brecce Bianche, 60131 Ancona, Italy; t.fioravanti@univpm.it (T.F.); a.splendiani@univpm.it (A.S.); t.righi@pm.univpm.it (T.R.)

2 Dipartimento di Biologia, Università degli Studi di Napoli Federico II, Via Cinthia 26, 80126 Napoli, Italy; nicomaio@unina.it

3 Dipartimento Scienze e Tecnologie Biologiche Chimiche e Farmaceutiche (STEBICEF), Università di Palermo, Via Archirafi 18, 90123 Palermo, Italy; sabrina.lobrutto@unipa.it

4 Struttura Diagnostica, Istituto Zooprofilattico Sperimentale della Puglia e della Basilicata (IZSPB), Via Manfredonia 20, 71121 Foggia, Italy; antonio.petrella@izspb.it

* Correspondence: v.caputo@staff.univpm.it

Received: 4 June 2020; Accepted: 22 June 2020; Published: 25 June 2020

Abstract: The Mediterranean monk seal (*Monachus monachus*) is one of the most endangered marine mammals in the world. This species has been threatened since ancient times by human activities and currently amounts to approximately 700 individuals distributed in the Eastern Mediterranean Sea (Aegean and Ionian Sea) and Eastern Atlantic Ocean (Cabo Blanco and Madeira). In other areas, where the species is considered "probably extinct", an increase in sporadic sightings has been recorded during recent years. Sightings and accidental catches of Mediterranean monk seals have become more frequent in the Adriatic Sea, mainly in Croatia but also along the coasts of Montenegro, Albania and Southern Italy. A Mediterranean monk seal pup was recovered on 27 January 2020 on the beach of Torre San Gennaro in Torchiarolo (Brindisi, Apulia, Italy). DNA was extracted from a tissue sample and the hypervariable region I (HVR1) of the mitochondrial DNA control region was amplified and sequenced. The alignment performed with seven previous published haplotypes showed that the individual belongs to the haplotype MM03, common in monk seals inhabiting the Greek islands of the Ionian Sea. This result indicates the Ionian Islands as the most probable geographical origin of the pup, highlighting the need to intensify research and conservation activities on this species even in areas where it seemed to be extinct.

Keywords: *Monachus monachus*; Mediterranean monk seal; mitochondrial DNA; Adriatic Sea; endangered species

---

1. Introduction

The Mediterranean monk seal (*Monachus monachus*, Hermann 1779) is the only living representative of the genus *Monachus* [1] and one of the most endangered mammals in the world [2]. It is a medium-sized phocid that usually inhabits waters up to 200 m in depth and is closely linked to coastal habitats for reproduction [3]. After mating in water, the females make their way to land and give birth to pups in coastal caves probably as an adaptation to predation, including hunting by humans [4,5]. Parturition usually occurs during the autumn season [6–8] but, in the Cabo Blanco colony, the birth of pups has also been observed in other months of the year as a result of favorable environmental conditions and the availability of food [6,9].

The Mediterranean monk seal was historically widespread in the Mediterranean Sea, Black Sea and Eastern Atlantic Ocean [3] but, at present the species consists of no more than 700 individuals inhabiting Ionian and Aegean waters within the Eastern Mediterranean area, the Madeira archipelago, and Cabo Blanco waters in the Eastern Atlantic [2]. The size of populations and the geographical distribution of the species have over time been impacted by several threats such as habitat deterioration, intentional or accidental killing and unusual mass mortality events [3]. Since prehistoric times *M. monachus* has been hunted by man [10,11], although intense exploitation probably occurred in the Roman and Medieval periods when monk seals were extensively hunted for their skin, oil, and meat [4,5]. Fishing is still the main cause of death of Mediterranean monk seals as many of them, considered responsible for a negative impact on fishing activities, are intentionally killed by fishermen [12] or accidentally become entangled in fishing nets (bycatch) [13]. In addition, coastal habitat degradation and increased tourism have reduced the mainland areas available for resting and pupping, putting at risk the reproductive capacity of the species [3]. In 1997, a mass mortality event was documented in Cabo Blanco highlighting the fact that, together with other stochastic events [3], viruses and toxic algal blooms [14,15] could also be important additional threats to monk seal survival.

In order to assess the conservation status of *M. monachus*, several studies have analyzed the genetic diversity levels of extant populations using nuclear and mitochondrial DNA sequences [16–19]. Both markers have highlighted very low levels of genetic diversity in Mediterranean monk seal populations, probably as a direct consequence of bottleneck events that occurred in the past [16–19]. In addition, the analysis of the most variable part of the mitochondrial DNA (mtDNA) control region has allowed the identification of three distinct sub-populations (Aegean Sea, Ionian Sea and Eastern Atlantic Ocean) characterised by different haplotypes [18]. Subsequently, the analysis of the same mtDNA marker, in both historical and current monk seal specimens, has allowed a re-evaluation of the previous result [19]. Gaubert et al. [19] showed that some mitochondrial haplotypes had a wider geographical distribution in the past and that, in agreement with previous results obtained by nuclear markers [17], there was a gene flow between populations which today seem to be genetically separate [19].

In recent years, a recovery in monk seal populations has been observed and the species has been reclassified as "Endangered" by the International Units for the Conservation of Nature (IUCN) [2]. Probably following the recovery of *M. monachus* populations, an increase in sightings has been observed in areas where the species seemed to be extinct, including North African countries, the coasts of the Levantine Sea, and those of the Adriatic-Ionian region [20]. Sightings of monk seals within the Adriatic basin are now numerous in Croatia, although vagrant individuals are also recorded in Montenegro, Albania and Apulia (Southern Italy) [20]. The presence of the Mediterranean monk seal along the Apulian coasts is not new; bone remains excavated from the Grotta Romanelli indicate that this area was a habitat for *M. monachus* individuals dating back to the Late Upper Paleolithic [21]. Over the centuries the species has progressively disappeared from Apulia, although some sightings were recorded in the Tremiti Island archipelago and in Salento in the second half of the 1900s [22] and between 2000 and the present day [20,23], indicating a probable recolonisation of this area.

On 27 January 2020, a Mediterranean monk seal pup was recovered stranded on the beach of Torre San Gennaro in Torchiarolo (Brindisi, Apulia, Italy). The young seal was debilitated and died the following day despite rescue efforts. A tissue sample was obtained, and a genetic analysis was performed in order to identify the most likely geographical origin of this monk seal individual.

2. Materials and Methods

The Mediterranean monk seal analyzed in this study (Figure 1) was recovered on the beach of Torre San Gennaro in Torchiarolo (Brindisi, Apulia, Italy) (Figure 2). After the death of the monk seal, standard measurements were taken and a necropsy was carried out by the "Istituto Zooprofilattico Sperimentale della Puglia e della Basilicata" (IZSPB, Foggia, Italy) in order to reveal the possible cause of death for the individual. During this procedure a piece of muscle tissue was sampled, stored in absolute ethanol and sent to the Laboratory of Evolutionary Biology (Università Politecnica delle Marche, Ancona, Italy) for genetic analysis.

Figure 1. A picture of the Mediterranean monk seal pup taken before the necropsy.

Figure 2. Map of the Adriatic-Ionian region showing the haplotype distribution of the Mediterranean monk seals previously analyzed. Only data with correct geographic locations are shown. The circle indicates data from Karamanlidis et al. [18], the triangle indicates data from Gaubert et al. [19], and the star indicates the location where our sample was found.

Genomic DNA was extracted from a small slice of muscle tissue using an automated nucleic acids extractor, MagCore® HF16 (RBC Bioscience Corp., Taipei, Taiwan), with the MagCore® Genomic DNA Tissue Kit (cartridge code 401) (RBC Bioscience Corp., Taipei, Taiwan) and following manufacturer's instructions. After the DNA extraction, the hypervariable region I (HVR1) of the mtDNA control region was amplified using primer pairs designed by Karamanlidis et al. [18].

Polymerase Chain Reaction (PCR) was performed in 25 µL volume containing: 5 µL of PrimeSTAR® GXL Polymerase Buffer (5×) (Takara, Shiga, Japan), 2 µL of dNTP mix (10 mM), 2 µL of Forward and Reverse Primers mix (5 µM), 0.8 µL of PrimeSTAR® GXL DNA Polymerase (1.25 U/µL) (Takara, Shiga, Japan), 3 µL of DNA template (40 ng/µL) and 12.2 µL of ultrapure sterile water. The amplification was carried out in a BioRad T100™ Thermal Cycler (BioRad, Hercules, CA, USA), using the following conditions: an initial denaturation step of 5 min at 95 °C, followed by 35 cycles of 30 s at 95 °C (denaturation), 30 s at 57 °C (annealing), 60 s at 72 °C (extension), and a final extension step of 7 min at 72 °C. The PCR product was run on a 2% agarose gel stained with GelRed™ (Biotium Inc., Hayward, CA, USA) to check the effectiveness of the amplification reaction and then sent to BMR Genomics (Padua, Italy) for Sanger sequencing. There, it was purified using exoSAP-IT™ (USB Corp., Cleveland, OH, USA) and sequenced in both directions on an ABIPRISM 3730XL automated sequencer (Applied Biosystems, Tokyo, Japan).

The sequence obtained was aligned on CLUSTALW [24] with those of the seven *M. monachus* haplotypes identified within the whole species range (GenBank Accession numbers in Table 1) [18,19]. The alignment was checked on BioEdit [25] and then a median-joining network (with $\varepsilon = 0$) [26] was designed using Network 10 software (Fluxus Technology Ltd., Colchester, UK, www.fluxus-engineering.com) in order to visualize the relationships between all haplotypes. In addition, with the aim of highlighting the geographical origin of the monk seal pup, information about the frequency and the geographical provenance of all haplotypes was obtained from previous studies [18,19] and added during the network construction. The information regarding the geographical origin of the haplotypes previously identified in the Adriatic-Ionian region [18,19] has also been included in a map.

Table 1. Diagnostic sites obtained from the alignment of the seven Mediterranean monk seal haplotypes. The haplotype belonging to the individual analyzed is highlighted in grey. Haplotype MM01 was used as a reference sequence. All identical nucleotides in other sequences are indicated as full stops.

Haplotype ID	GenBank Accession No.	Diagnostic Sites					
		105	114	118	248	279	508
MM01	KT935311	A	A	A	A	C	G
MM02	KT935307	G	.	.	.	.	.
MM03	KT935310	.	.	.	.	.	A
MMBR	MT524708	.	.	.	.	.	A
MM05	KT935309	.	.	G	.	.	.
MM06	MG570470	.	.	G	.	T	.
MM04	KT935308	.	.	G	.	.	A
MM07	MG570469	.	G	G	G	.	A

Capital letters indicate nucleotides: A—adenine, C—cytosine, G—guanine, T—thymine.

3. Results

The Mediterranean monk seal was a female weighing 22.5 kg and 118 cm long. Pelage color, weight and standard length suggest that it was a pup of about 2–4 months of age. Genetic analysis of the individual allowed us to successfully amplify and sequence a 524 bp fragment of the HVR1 of the mtDNA control region. The sequence obtained was stored in the GenBank repository with the Accession number MT524708. After the alignment of this sequence with those of the seven *M. monachus* haplotypes known so far, six diagnostic sites were identified, allowing the assignment of the pup sequence to the haplotype MM03 (Table 1). The map and the haplotype network, which include information about geographical origin of the known haplotypes, showed that the haplotype MM03 has been previously observed only in Greek populations inhabiting the Ionian Sea and in a sample from the Adriatic Sea (Figures 2 and 3).

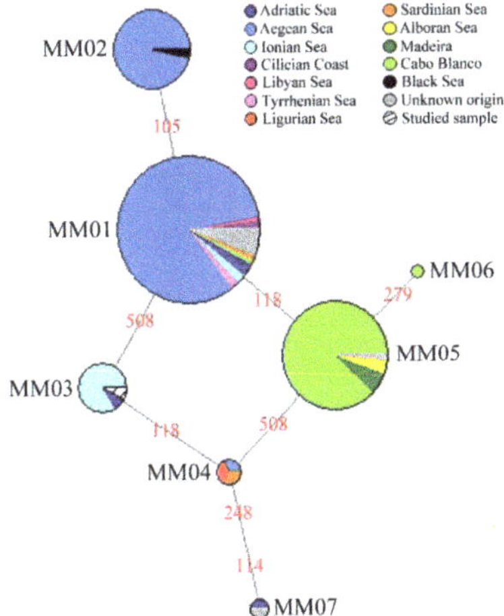

Figure 3. Median-joining network of the seven Mediterranean monk seal haplotypes. Colors indicate geographic sampling locations. Size of each circle is proportional to the haplotype frequency. Numbers in red indicate mutated positions.

4. Discussion

A Mediterranean monk seal found along the Apulian coast (Southern Italy) at the end of January 2020 was genetically analyzed in this study. The amplification and sequencing of the HVR1 fragment of the mtDNA control region have allowed the haplotype identification of the monk seal. The individual studied belonged to the haplotype MM03, one of the least common within the current distribution range of the species [18,19]. The haplotype MM03 was previously found only in 10 individuals sampled from the Greek islands (Kefalonia and Zakynthos) in the Ionian Sea [18] and from one individual in the Adriatic Sea [19]. The latter was a museum specimen which had been caught along the west coast of Brac (Croatia) in 1914 [19]. Because of the absence of reproductive populations of the species, the Mediterranean monk seal is currently considered "probably extinct" within the Adriatic Sea and the few individuals recorded in this basin are usually considered as "vagrants", individuals coming from areas where the species is more abundant and reproductively active [20,23]. While the Adriatic waters are only sporadically inhabited by Mediterranean monk seals, the Ionian Islands still host several *M. monachus* individuals. In fact, one of the largest populations of this species is found in the Eastern Mediterranean Sea and approximately 300–400 individuals are distributed along the Greek coasts [3]. The Ionian Islands are important sites for the conservation of the Mediterranean monk seal because they are characterised by a heterogeneous coastal habitat, which includes both open sandy beaches and several rocky areas, with marine caves suitable as resting and breeding grounds for monk seals [27,28]. Studies on the behavior and migratory capacity of the Mediterranean monk seal are extremely rare due to the difficulty in performing tagging and tracking experiments [29]. However, observations carried out in different areas in Greece have highlighted the ability of sub-adults and adults to cover a total distance of 100–300 km, with a daily mean of 10–40 km [29]. Moreover, juvenile individuals have also been observed in areas hundreds of kilometers away from their putative origin [23,30,31]. One juvenile female of approximately six months old was found dead in a fishing net along the Libyan coast. The genetic analysis carried out showed a probable origin in the Eastern Mediterranean Sea, suggesting that the young individual may have travelled hundreds of kilometers before being captured [31]. The observation of the haplotype MM03 in the studied specimen and the distance of this area from the Apulian coast (ca. 320 km), allows us to hypothesise the most probable origin of the individual analyzed to be from the Greek islands in the Ionian Sea. However, we cannot exclude the possibility that it was born in Apulia.

The morphological features and the small size of the *M. monachus* individual suggest that it was a very young specimen. Even if the standard length of 118 cm places the pup in age class category three which includes 1–1.7-month-old individuals [32], the morphology allows us to classify it in age class category four (2–4 month old pups) [32]. The monk seal seems to have completed the molt, so it had shed the "lanugo" (a dark woolly coat typical of newborns), and presented a grey color on its back and a pale belly [32,33]. The estimated age of about 2–4 months could suggest that the pup was born along the Apulian coast in the autumn. This hypothesis is supported by the presence in this area of several habitats suitable for the reproduction of monk seals [34] and by the fact that at only 2–4 months of age the pup would still be breastfed by the mother. Mediterranean monk seal females give birth to only one pup per year using coastal caves as shelter. Parturition usually occurs from September to November [6–8] and the pup is breastfed up to five months of age, after which it begins to move and feed itself [35,36]. The great ability of monk seals to travel long distances in the open sea and the genetic results probably indicate that the mother of the pup was a vagrant individual coming from the Ionian Islands (Greece) who chose the Apulian coasts for parturition.

In conclusion, the analysis performed in this study shows that the Mediterranean monk seal pup found along the Apulian coast was probably born in this area and that its mother came from the Greek islands in the Ionian Sea. An origin from the Ionian Islands has been hypothesized as they host the only reproductively active population of monk seals within the Adriatic-Ionian region. Other areas of the Adriatic Sea are only sporadically frequented by individuals of this species. The finding of a Mediterranean monk seal born along the Apulian coast is an important event that proves the recovery of this species in the Mediterranean basin and the probable recolonisation of areas from which it had disappeared. The phenomenon of recolonization has already been hypothesized taking into consideration the increase in sightings in areas of the Mediterranean Sea where the species is not a stable resident [20]. For example, along the Lebanese coast a total of 47 sightings of Mediterranean monk seals was recorded from 2004 to April 2020. These numerous and recent sightings could indicate the presence of remaining individuals from an ancient population or could suggest a probable recolonisation of the Lebanese coast by individuals from the closest reproductive populations of Turkey and Cyprus, indicating an attempt to expand their distribution range [37]. On the other hand, the lack of knowledge about the life cycle, behavior, and real distribution of the species highlights the need to intensify research and monitoring activities. Gathering new information on *M. monachus* will allow us to evaluate the current distribution of the monk seals within the Mediterranean Sea and to understand if there are suitable habitats for the recolonisation of the species and the establishment of new reproductive colonies. In addition, in areas where *M. monachus* populations are already present and in those suitable for recolonisation, fishing activities could be well regulated and the degree of anthropization and coastal habitat deterioration could be reduced. The development of appropriate management plans will be therefore useful to reduce the threats to the species and to promote the recovery and conservation of Mediterranean monk seal within its original distribution range.

Author Contributions: Conceptualization, T.F. and V.C.B.; methodology, T.F. and A.S.; formal analysis, T.F. and A.P.; investigation, T.F.; data curation, T.F.; writing—original draft preparation, T.F.; writing—review and editing, A.S., T.R., N.M., S.L.B., A.P. and V.C.B.; supervision, V.C.B.; project administration, V.C.B.; funding acquisition, V.C.B. All authors have read and agreed to the published version of the manuscript.

Funding: This research was funded by "Università Politecnica delle Marche", grant number 040017_R.SCIENT.A_2019_CAPUTO_BARUCCHI_V.

Acknowledgments: We are grateful to Frances Marie Baker (Centro di Supporto per l'Apprendimento delle Lingue, CSAL, Università Politecnica delle Marche) for the English Language editing and to Prof. Nicola Zizzo (Dipartimento Di Medicina Veterinaria, Università degli Studi di Bari Aldo Moro) for the useful collaboration. An ethics statement is not necessary for this work as the Mediterranean monk seal analyzed died of natural causes and the sample used for the genetic analysis was provided by the institute that carried out the authorized necropsy of the individual (Istituto Zooprofilattico Sperimentale della Puglia e della Basilicata, IZSPB).

Conflicts of Interest: The authors declare no conflict of interest.

References

1. Scheel, D.M.; Slater, G.J.; Kolokotronis, S.O.; Potter, C.W.; Rotstein, D.S.; Tsangaras, K.; Greenwood, A.D.; Helgen, K.M. Biogeography and taxonomy of extinct and endangered monk seals illuminated by ancient DNA and skull morphology. *Zookeys* **2014**, *409*, 1–33. [CrossRef]

2. Karamanlidis, A.A.; Dendrinos, P. *Monachus monachus. IUCN Red List Threat. Species 2015*; International Union for Conservation of Nature and Natural Resources: Gland, Switzerland, 2015. [CrossRef]

3. Karamanlidis, A.A.; Dendrinos, P.; de Larrinoa, P.F.; Gücü, A.C.; Johnson, W.M.; Kiraç, C.O.; Pires, R. The Mediterranean monk seal *Monachus monachus*: Status, biology, threats, and conservation priorities. *Mamm. Rev.* **2016**, *46*, 92–105. [CrossRef]

4. Johnson, W.M. Monk seals in post-classical history. The role of the Mediterranean monk seal (*Monachus monachus*) in European history and culture, from the fall of Rome to the 20th century. In *Mededelingen No. 39*; Netherlands Commission for International Nature Protection: Leiden, The Netherlands, 2004; pp. 1–91.

5. Johnson, W.M.; Lavigne, D.M. Monk seals in antiquity. The Mediterranean monk seal (*Monachus monachus*) in ancient history and literature. In *Mededelingen No. 35*; Netherlands Commission for International Nature Protection: Leiden, The Netherlands, 1999.

6. Pastor, T.; Aguilar, A. Reproductive cycle of the female Mediterranean monk seal in the Western Sahara. *Mar. Mamm. Sci.* **2003**, *19*, 318–330. [CrossRef]

7. Gucu, A.C.; Gucu, G.; Orek, H. Habitat use and preliminary demographic evaluation of the critically endangered Mediterranean monk seal (*Monachus monachus*) in the Cilician Basin (Eastern Mediterranean). *Biol. Conserv.* **2004**, *116*, 417–431. [CrossRef]

8. Karamanlidis, A.A.; Paravas, V.; Trillmich, F.; Dendrinos, P. First observations of parturition and postpartum behavior in the Mediterranean monk seal (*Monachus monachus*) in the eastern Mediterranean. *Aquat. Mamm.* **2010**, *36*, 27–32. [CrossRef]

9. Gazo, M.; Layna, J.F.; Aparicio, F.; Cedenilla, M.A.; González, L.M.; Aguilar, A. Pupping season, perinatal sex ratio and natality rates of the Mediterranean monk seal (*Monachus monachus*) from the Cabo Blanco colony. *J. Zool.* **1999**, *249*, 393–401. [CrossRef]

10. Stringer, C.B.; Finlayson, J.C.; Barton, R.N.E.; Fernández-Jalvo, Y.; Cáceres, I.; Sabin, R.C.; Rhodes, E.J.; Currant, A.P.; Rodríguez-Vidal, J.; Giles-Pacheco, F.; et al. Neanderthal exploitation of marine mammals in Gibraltar. *Proc. Natl. Acad. Sci. USA* **2008**, *105*, 14319–14324. [CrossRef] [PubMed]

11. Morales-Pérez, J.V.; Pérez Ripoll, M.; Jordá Pardo, J.F.; Álvarez-Fernández, E.; Maestro González, A.; Aura Tortosa, J.E. Mediterranean monk seal hunting in the regional Epipalaeolithic of Southern Iberia. A study of the Nerja Cave site (Málaga, Spain). *Quat. Int.* **2019**, *515*, 80–91. [CrossRef]

12. Androukaki, E.; Adamantopoulou, S.; Dendrinos, P.; Tounta, E.; Kotomatas, S. Causes of mortality in the Mediterranean monk seal (*Monachus monachus*) in Greece. *Contrib. Zool. Ecol. East. Mediterr. Reg.* **1999**, *1*, 405–411.

13. Karamanlidis, A.A.; Androukaki, E.; Adamantopoulou, S.; Chatzispyrou, A.; Johnson, W.M.; Kotomatas, S.; Papadopoulos, A.; Paravas, V.; Paximadis, G.; Pires, R.; et al. Assessing accidental entanglement as a threat to the Mediterranean monk seal *Monachus monachus. Endanger. Species Res.* **2008**, *5*, 205–213. [CrossRef]

14. Osterhaus, A.; Van De Bildt, M.; Vedder, L.; Martina, B.; Niesters, H.; Vos, J.; Van Egmond, H.; Liem, D.; Baumann, R.; Androukaki, E.; et al. Monk seal mortality: Virus or toxin? *Vaccine* **1998**, *16*, 979–981. [CrossRef]

15. Reyero, M.; Cacho, E.; Martínez, A.; Vázquez, J.; Marina, A.; Fraga, S.; Franco, D.J.M. Evidence of saxitoxin derivatives as causative agents in the 1997 mass mortality of monk seals in the Cape Blanc peninsula. *Nat. Toxins* **1999**, *7*, 311–315. [CrossRef]

16. Pastor, T.; Garza, J.C.; Allen, P.; Amos, W.; Aguilar, A. Low genetic variability in the highly endangered mediterranean monk seal. *J. Hered.* **2004**, *95*, 291–300. [CrossRef] [PubMed]

17. Pastor, T.; Garza, J.C.; Aguilar, A.; Tounta, E.; Androukaki, E. Genetic diversity and differentiation between the two remaining populations of the critically endangered Mediterranean monk seal. *Anim. Conserv.* **2007**, *10*, 461–469. [CrossRef]

18. Karamanlidis, A.A.; Gaughran, S.; Aguilar, A.; Dendrinos, P.; Huber, D.; Pires, R.; Schultz, J.; Skrbinšek, T.; Amato, G. Shaping species conservation strategies using mtDNA analysis: The case of the elusive Mediterranean monk seal (*Monachus monachus*). *Biol. Conserv.* **2016**, *193*, 71–79. [CrossRef]

19. Gaubert, P.; Justy, F.; Mo, G.; Aguilar, A.; Danyer, E.; Borrell, A.; Dendrinos, P.; Öztürk, B.; Improta, R.; Tonay, A.M.; et al. Insights from 180 years of mitochondrial variability in the endangered Mediterranean monk seal (*Monachus monachus*). *Mar. Mamm. Sci.* **2019**, *35*, 1489–1511. [CrossRef]

20. Bundone, L.; Panou, A.; Molinaroli, E. On sightings of (vagrant?) monk seals, *Monachus monachus*, in the Mediterranean Basin and their importance for the conservation of the species. *Aquat. Conserv. Mar. Freshw. Ecosyst.* **2019**, *29*, 554–563. [CrossRef]

21. Cassoli, P.F.; Fiore, I.; Tagliacozzo, A. Butchery and exploitation of large mammals in the Epigravettian levels of Grotta Romanelli. *Anthropozoologica* **1997**, *25*, 309–318.

22. Di Turo, P. Presenza della foca monaca (*Monachus monachus*) nell'area mediterranea con particolare riferimento alla Puglia. *Thalassia Salentina* **1984**, *14*, 66–84. [CrossRef]

23. Mo, G. Mediterranean monk seal (*Monachus monachus*) sightings in Italy (1998–2010) and implications for conservation. *Aquat. Mamm.* **2011**, *37*, 236–240. [CrossRef]

24. Larkin, M.A.; Blackshields, G.; Brown, N.P.; Chenna, R.; Mcgettigan, P.A.; McWilliam, H.; Valentin, F.; Wallace, I.M.; Wilm, A.; Lopez, R.; et al. Clustal W and Clustal X version 2.0. *Bioinformatics* **2007**, *23*, 2947–2948. [CrossRef] [PubMed]

25. Hall, T.A. BioEdit: A user-friendly biological sequence alignment editor and analysis program for Windows 95/98/NT. *Nucleic Acids Symp. Ser.* **1999**, *41*, 95–98.

26. Bandelt, H.-J.; Forster, P.; Röhl, A. Median-joining networks for inferring intraspecific phylogenies. *Mol. Biol. Evol.* **1999**, *16*, 37–48. [CrossRef] [PubMed]

27. Panou, A.; Jacobs, J.; Panos, D. The endangered mediterranean monk seal *Monachus monachus* in the Ionian Sea, Greece. *Biol. Conserv.* **1993**, *64*, 129–140. [CrossRef]

28. Panou, A. Monk seal sightings in the central Ionian Sea. A network of fishermen for the protection of the marine resources. In Proceedings of the European Cetacean Society Annual Conference, Istanbul, Turkey, 28 February 2009; p. 6.

29. Adamantopoulou, S.; Androukaki, E.; Dendrinos, P.; Kotomatas, S.; Paravas, V.; Psaradellis, M.; Tounta, E.; Karamanlidis, A.A. Movements of Mediterranean monk seals (*Monachus monachus*) in the Eastern Mediterranean Sea. *Aquat. Mamm.* **2011**, *37*, 256–261. [CrossRef]

30. Bayed, A. Further observations of Mediterranean monk seals on the north Atlantic coast of Morocco. *Monachus Guard.* **2001**, *4*, 45–47.

31. Alfaghi, I.A.; Abed, A.S.; Dendrinos, P.; Psaradellis, M.; Karamanlidis, A.A. First confirmed sighting of the mediterranean monk seal (*Monachus monachus*) in Libya since 1972. *Aquat. Mamm.* **2013**, *39*, 81–84. [CrossRef]

32. Mo, G.; Zotti, A.; Agnesi, S.; Finola, M.G.; Bernardini, D.; Cozzi, B. Age classes and sex differences in the skull of the Mediterranean monk seal, *Monachus monachus* (Hermann, 1779). A study based on bone shape and density. *Anat. Rec.* **2009**, *292*, 544–556. [CrossRef]

33. Samaranch, R.; González, L.M. Changes in morphology with age in Mediterranean monk seals (*Monachus monachus*). *Mar. Mamm. Sci.* **2000**, *16*, 141–157. [CrossRef]

34. Bundone, L.; Fai, S.; D'Ambrosio, P.; Onorato, R.; Minonne, F.; Molinaroli, E. Coastal habitat availability for the Mediterranean monk seal, *Monachus monachus* (Hermann, 1779), in Salento: Preliminary results. *Biol. Mar. Mediterr.* **2014**, *21*, 253–254.

35. Aguilar, A.; Cappozzo, L.H.; Gazo, M.; Pastor, T.; Forcada, J.; Grau, E. Lactation and mother-pup behaviour in the Mediterranean monk seal *Monachus monachus*: An unusual pattern for a phocid. *J. Mar. Biol. Assoc. UK* **2007**, *87*, 93–99. [CrossRef]

36. Kıraç, C.O.; Ok, M. Diet of a Mediterranean monk seal *Monachus monachus* in a transitional post-weaning phase and its implications for the conservation of the species. *Endanger. Species Res.* **2019**, *39*, 315–320. [CrossRef]

37. SPA/RAC-UNEP/MAP. *On the Occurrence of the Mediterranean Monk Seal* Monachus monachus *(Hermann, 1779) in the Lebanese Waters (Eastern Mediterranean Sea)*; Badreddine, A., Limam, A., Ben-Nakhla, L., Eds.; SPA/RAC: Tunis, Tunisia, 2020; p. 12.

Article

Loss of Mitochondrial Genetic Diversity in Overexploited Mediterranean Swordfish (*Xiphias gladius*, 1759) Population

Tommaso Righi, Andrea Splendiani, Tatiana Fioravanti, Elia Casoni, Giorgia Gioacchini, Oliana Carnevali and Vincenzo Caputo Barucchi *

Dipartimento di Scienze della Vita e dell'Ambiente, Università Politecnica delle Marche, Via Brecce Bianche, 60131 Ancona, Italy; t.righi@pm.univpm.it (T.R.); andreasplendiani@hotmail.com (A.S.); t.fioravanti@pm.univpm.it (T.F.); eliacasoni@libero.it (E.C.); giorgia.gioacchini@staff.univpm.it (G.G.); o.carnevali@staff.univpm.it (O.C.)
* Correspondence: v.caputo@staff.univpm.it

Received: 4 March 2020; Accepted: 24 April 2020; Published: 26 April 2020

Abstract: Intense and prolonged mortality caused by over-exploitation could drive the decay of genetic diversity which may lead to decrease species' resilience to environmental changes, thus increasing their extinction risk. Swordfish is a high commercial value species, especially in the Mediterranean Sea, where it is affected by high catch levels. Mediterranean swordfish consist of a population genetically and biologically distinct from Atlantic ones and therefore managed as a separate stock. The last Mediterranean swordfish stock assessment reported that in the last forty years Mediterranean swordfish has been overfished and, to date, it is still subject to overfishing. A comparison between an available mitochondrial sequence dataset and a homologous current sample was carried out to investigate temporal genetic variation in the Mediterranean swordfish population over near twenty years. Our study provides the first direct measure of reduced genetic diversity for Mediterranean swordfish during a short period, as measured both in the direct loss of mitochondrial haplotypes and reduction in haplotype diversity. A reduction of the relative females' effective population size in the recent sample has been also detected. The possible relationship between fishery activities and the loss of genetic diversity in the Mediterranean swordfish population is discussed.

Keywords: Mediterranean Sea; mtDNA; control region; swordfish

---

1. Introduction

The swordfish (*Xiphias gladius*) is a large pelagic and migratory fish found in the open waters of all oceans, including the Mediterranean Sea [1]. Swordfish migration is complex and multi-directional [1]. Seasonally adults swordfish migrate for reproductive reasons to spawning grounds, while a diel pattern in vertical movement occurring for feeding, reaching deep water (300–600 m) during daylight and staying closer to the sea-surface during the night [2,3]. Swordfish growth is relatively rapid and sexually dimorphic. Females generally reach a larger maximum size than males and large individuals are usually female [2,4]. Mediterranean swordfish consist of a population genetically and biologically distinct from the Atlantic one. The growth parameters are different, and the sexual maturity is reached at younger ages than in the Atlantic [4]. In the Mediterranean, the estimated size at which 50% (L_{50}) of the female population is mature occurs from 131.5 cm [5] and 140 cm [6], corresponding to 3–4-year-old fish [2]. Males reach sexual maturity at smaller sizes and mature specimens have been found at about 90 cm (Lower jaw fork length, LJFL) [7]. Genetic differentiation [8], as well as several differences in biological parameters [2], have been used to identify and manage the Atlantic and Mediterranean swordfish as separated stocks [7]. A popular genetic marker used to examine swordfish populations is the mitochondrial control region (D-loop). As reported for other large pelagic species (*Thunnus alalunga* [9]; *Thunnus thynnus* [10] *Seriola dumerili* [11]) the mitochondrial control region of swordfish can be divided into two divergent clades. Specifically for swordfish, a ubiquitous Clade I widespread in all stocks, and Clade II which decreases in abundance with distance from the Mediterranean Sea [10,12,13] and is absent from the Pacific Ocean [14–16]. These phylogeographic associations support the inferred pattern of unidirectional historical gene flow from the Indo–Pacific into the Atlantic and the Atlantic origin of Clade II [10]. A subdivision in two monophyletic subclades to the Atlantic and Mediterranean Clade II lineage confirmed genetic isolation between the two stocks [10,13]. Furthermore, within the Mediterranean Sea Swordfish population sub-structuring was suggested based on a clinal decrease in mitochondrial DNA control region intra-clade genetic variability from Western and the Eastern basins [17]. The Mediterranean population showed the lowest levels of genetic variation compared to any other population worldwide [10,12,13,18,19]. According to Bremer et al. [10], the reduced level of variation in Mediterranean swordfish mtDNA is congruent with multiple faunal extinctions and subsequent re-colonization events occurred during the Pleistocene. However, according to Kotoulas et al. [20], the low genetic variability observed using nuclear markers could be related to the small effective population size of Mediterranean swordfish, probably caused by fishing pressure [19]. Moreover, according to Arocha [4], prolonged exploitation of swordfish in the Mediterranean area may have reduced the numbers of older specimens and consequently explain the lower mean size of the catches and lower size of maturation observed for the Mediterranean population.

Swordfish is a high commercial value species, especially in the Mediterranean Sea, where despite this basin represents less than 10% of the swordfish global range, catch levels are relatively high and similar to those of larger areas such as the North Atlantic [21]. Some characteristics as the presence of a larger spawning area in relation to the area of distribution of the stock, lower abundance of predators and higher recruitment were suggested as the determining factors explaining the higher abundance of swordfish in the Mediterranean Sea. However, in the last forty years, the swordfish stock of the Mediterranean Sea has been overfished and, to date, is still subject to overfishing [7]. The Spawning Stock Biomass (SSB) estimated represents less than 15% of the maximum sustainable yield (B_{MSY}) and the fish mortality caused by harvesting (F) is almost twice the maximum sustainable yield. The 50–70% of the total yearly catches is represented by fish of small size often less than 3 years old, with a high level of immature swordfish reported [7]. The loss of genetic variation caused by population reduction is an important concern in fisheries management. Intense and prolonged mortality caused by overexploitation could drive the decay of genetic diversity across a wide range of marine fishes [22]. Genetic diversity has been defined as the variety of alleles and genotypes present in a population and that is reflected in morphological, physiological and behavioral differences between individuals and populations [23]. The reduction of genetic diversity may lead to decreased species resilience to environmental changes, thus increasing their extinction risk [23,24]. Small and isolated populations are more vulnerable to genetic degradation [23]. Moreover, in marine fishes, effective population size (Ne) can be several orders of magnitude smaller than census size (N) [25–27] and species may suffer a loss of genetic diversity under fishing pressure despite large census size.

The most powerful way to detect loss of genetic variation in exploited populations is the examination of genetic samples collected over time [28]. In this study, a dataset of available mitochondrial sequences of Mediterranean swordfish sampled in the 1990s and early 2000s has been compared with homologous mtDNA data obtained from current samples (2016–2018). This comparison was carried out to investigate if a temporal genetic variation occurred in the Mediterranean swordfish, where this species is considered threatened [21] and overexploited.

2. Materials and Methods

2.1. Sampling and DNA Extraction

A total of 287 swordfish were collected from six areas within the eastern, central and western Mediterranean regions, trying to obtain a representative coverage of the basin (For details see Table 1 and Figure 1).

Table 1. Sampling details for swordfish analyzed in this study.

Sampling Area (FAO Fishing Area/Geographical Subarea)	Sampling Date	n	Mean LJFL (cm) (s.d.)
Balearic Sea (GSA 5, 6)	July 2016–August 2016–September 2016 September 2019	84	122.1 (22,2)
Southern Sicily (GSA 15, 16)	July 2016 June 2017–July 2017 June 2018	59	133.7 (32.0)
Aegean Sea (GSA 22)	August 2016	17	124.2 (21.0)
South Adriatic Sea (GSA 18)	September 2016	62	113.5 (16.6)
Tyrrhenian Sea (GSA 10)	May 2017	16	120.8 (23.8)
Sardinian Sea (GSA 11.2)	June 2017–October 2017 May 2018–Jun 2018	49	125.0 (32.6)
Total		287	

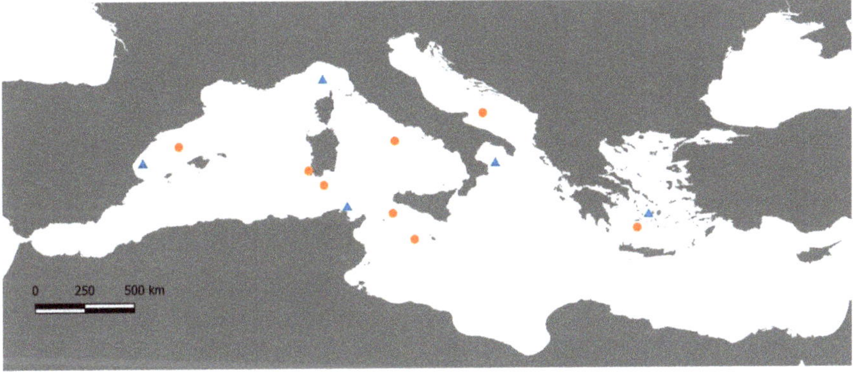

Figure 1. Map of the Mediterranean Sea with the Swordfish sampling sites. Blue triangles represent historical sampling localities from Viñas et al. [17] and orange dots represent sampling locations of current Swordfish samples.

Samples were achieved at the fishing landing of commercial catch by longline or trap bycatch (only in case of Sardinian samples) from May to October in three years 2016–2018. For each individual, a piece of the caudal fin was collected and was stored in ethanol absolute and kept at −20 °C until DNA extraction. Total genomic DNA was extracted using specific cartridge 401 in the MagCore® automated Nucleic Acid extractor (RBC Bioscience, New Taipei City, Taiwan) following the specific protocol. Historical mitochondrial data were obtained from a re-analysis of the dataset available from Viñas [17] (Table S2). That work included specimens sampled in several Mediterranean areas (Figure 1) from 1992 to 1995 and the early 2000s (see [11] for sample details), allowing us to evaluate the mitochondrial genetic variability between two complementary sample groups of Mediterranean swordfish over a 20-year period.

2.2. Mitochondrial DNA Amplification and Analysis

PCR-RFLP (Restrictin Fragment Length Polymorphism) analysis was performed to identify the two mitochondrial lineages. The restriction enzymes were selected by virtual restriction analysis (see Supplementary material S3 for virtual analysis details). A 360 bps long portion of the hypervariable L-domain of the mitochondrial control region was amplified using primers L15998 (5'-TAC CCC AAA CTC CCA AAG CTA-3') [12] and SWO 5' CCC TGT GAA ATA TGC TGG TTG 3' (designed in this study). The amplification was carried out in 25 µL reaction volume containing: 1 × MyTaq reaction buffer (BioLine GmbH, Luckenwalde, Germany), 10 pmol of each primer, 0.5 U of BIOTAQ DNA-Polymerase (BioLine GmbH, Luckenwalde, Germany) and approximately 80 ng of the isolated DNA. For thermal cycler Viñas et al. [9] has been followed. Amplicons were double-digested with selected PacI and VspI restriction enzymes (Thermo Fisher Scientific, Waltham, Massachusetts, MA, USA). The digestion mix reaction was carried out in a total volume of 20µL with 4 µL of amplified DNA, 2 µL of Buffer G 10X (Thermo Fisher Scientific, Waltham, Massachusetts, MA, USA), and 0.3 µL [10 U/µL] of each enzyme. The digestion took place at 37 °C overnight. Restriction profiles were visualized in 2% agarose gel. Subsequently, to screen mitochondrial genetic variability a PCR-SSCP (Single Strand Conformation Polimorphism) analysis was performed. 5 µL amplicon was added to 4 µL loading buffer (98% formamide, 10 mM, EDTA (0.5 M, pH 8), 0.05% bromophenol blue, 0.05% xylene cyanol), heated to 95 °C for 5 min and immediately chilled on ice. Vertical electrophoresis run was performed in a nondenaturing polyacrylamide gel (8% acrylamide/polyacrylamide [49:1], 10% glycerol) at 5 W for 12 h at room temperature, with 1 × TBE as the running buffer. To identify different haplotypes Sanger sequencing of *D-loop* was performed on a subsample for each SSCP pattern on an ABI 3730XL DNA (Applied Biosystems, Foster City, CA, USA). To validate the sensitivity of SSCP technique in the identification of different haplotypes, all individuals belonging to the Clade II lineage, which showed a fewer number of SSCP morphs, were sequenced. Sequences were aligned using ClustalW [29] and checked by eye in BioEdit [30]. Analysis of sequence variability was analyzed pooling all samples together and, following Viñas [17] grouping samples in Western (Balearic, Sardinian, Tyrrhenian) and Eastern (Sicilian, Adriatic and Aegean), according to the basin of origin. However, no genetic differentiation was detected by AMOVA (Analysis of Molecular VAriance) between the western and eastern group therefore we reported the only comparison among whole historical and contemporary datasets.

The number of haplotypes (N), and polymorphic site, haplotypic (h) and nucleotide diversities (π) were computed in Arlequin v. 3.5 [31]. The analysis was performed for both dataset pooling Clades I and II together, and for each clade separately. To account for the sampling variance of haplotypic richness, rarefaction curves [32] were generated for both samples. Rarefaction curves were generated using iNEXT [33,34]. Median-joining (MJ) networks for the entire dataset were built using the software Network v 4.510 [35] with the default settings.

The extent of genetic differentiation between historical and contemporary datasets was investigated with an analysis of molecular variance, AMOVA [36], and the exact tests of population differentiation implemented in Arlequin [31]. The AMOVA significance levels were determined by 10,000 permutations while the exact test was conducted by using 10,000 Markov chain steps and 1000 dememorization steps. Finally, temporal changes in the relative effective number of females (N_{ef}) was estimated comparing θ for each dataset. The θ is expected to be equal to $2\,N_{ef}\,\mu$ for neutral mutations in mtDNA, where μ is the mutation rate per generation and N_{ef} is the effective population size of females. Considering μ to be constant between two temporal samples, we can treat differences in θ among samples as differences in relative female effective population sizes. Population size parameter (θ) was inferred for each dataset using the maximum likelihood (ML) inference in LAMARC 2.1 [37] using Watterson's estimator as starting points, default model of evolution was used. A Markov Chain Monte Carlo was run for ten short chains and three long chains with 1000 and 10,000 recorded genealogies, respectively, after discarding the first 10,000 genealogies. One of every 20 reconstructed genealogies was sampled for both short and long chains.

3. Results

The SSCP analysis of 360 bps portion of mtDNA D-loop from 287 contemporary Mediterranean swordfish revealed 36 different morphs. For each morph, a subsample has been sequenced for a total of 172 swordfish, corresponding to 60% of the total sample. The sequencing of all individuals belonging to the Clade II corroborated the SSCP profile interpretation. The sequence length ranged from 290 to 297 bps due to the presence of 1–3 tandemly repeated copies of the motif TACA near the 5′ end. Sequence alignment revealed 49 polymorphic sites, defining 36 distinct haplotypes (Table 2).

Table 2. Variable sites in the control-region segment of the 36 haplotypes of swordfish mtDNA detected in the modern sample. Numbers above sites refer to nucleotide position on the L-strand of the control region of the swordfish. Identity with sequence 1 is indicated by dots and deletions of 1 residue with dashes.

Genbank Accession Number	2	9	12	17	18	19	20	21	22	23	24	46	60	67	77	83	91	92
AY650768	G	T	A	T	A	C	A	-	-	-	-	C	A	A	C	A	T	T
MN652597	.	.	.	.	.	.	.	T	A	C	A	.	.	.	.	.	.	.
AY650778	.	.	.	.	.	.	.	-	-	-	-	.	.	.	.	.	.	.
AY650860	.	.	.	.	.	.	.	-	-	-	-	.	.	.	.	.	.	.
MN652598	A	.	.	.	.	.	.	-	-	-	-	.	.	.	.	.	.	.
MN652599	.	.	.	.	.	.	.	-	-	-	-	.	.	.	.	.	.	.
EU827759	.	.	.	.	.	.	.	-	-	-	-	.	.	.	.	G	.	.
MN652600	.	.	.	.	.	.	.	T	A	C	A	.	.	.	.	.	.	.
MN652601	.	.	.	.	.	.	.	-	-	-	-	.	.	.	.	.	.	.
AY650836	.	.	.	.	.	.	.	-	-	-	-	.	.	.	.	.	.	.
AY650821	.	.	.	.	.	.	.	-	-	-	-	.	.	.	.	.	.	.
MN652602	.	.	.	.	.	.	.	-	-	-	-	.	.	.	.	.	.	.
AY650858	.	.	.	.	.	.	.	-	-	-	-	.	.	G	.	.	.	.
MN652603	.	.	.	.	.	.	.	-	-	-	-	.	.	.	.	.	.	.
MN652604	.	.	.	.	.	.	.	-	-	-	-	.	.	.	.	.	.	.
MN652605	.	.	.	.	.	.	.	-	-	-	-	.	.	.	.	.	.	.
MN652606	.	.	.	.	.	.	.	-	-	-	-	.	.	.	.	.	.	.
MN652607	.	.	.	.	.	.	.	-	-	-	-	.	.	.	T	.	.	.
MN652608	.	-	.	.	.	.	.	-	-	-	-	.	.	.	.	.	.	.
AY650781	.	.	.	.	.	.	.	-	-	-	-	.	.	.	.	.	.	.
MN652609	.	.	.	.	.	.	.	-	-	-	-	.	.	.	.	.	.	.
AY650763	.	.	.	.	.	.	.	-	-	-	-	.	.	.	T	.	.	.
MN652595	.	.	.	.	.	.	.	-	-	-	-	.	G	.	T	.	.	.
AY650855	.	.	.	.	.	.	.	T	A	C	A	.	.	.	T	.	.	.
MN652596	.	.	.	.	.	.	.	-	-	-	-	.	.	.	T	.	.	.
AY650861	.	.	.	.	.	.	.	-	-	-	-	.	.	.	T	.	.	.
AY650809	.	.	.	-	-	-	-	-	-	-	-	T	.	.	T	.	.	.
MN652610	.	.	.	-	-	-	-	-	-	-	-	T	.	.	T	.	.	.
AY650762	.	.	.	-	-	-	-	-	-	-	-	T	.	.	T	.	.	.
AY650814	.	.	.	-	-	-	-	-	-	-	-	T	.	.	T	.	.	.
AY650805	.	.	.	-	-	-	-	-	-	-	-	T	.	.	T	.	.	.
AY650761	.	.	.	-	-	-	-	-	-	-	-	T	.	.	T	.	C	.
AY650829	.	.	.	-	-	-	-	-	-	-	-	T	.	.	T	.	.	.
EU827771	.	.	.	-	-	-	-	-	-	-	-	T	.	.	T	.	.	.
MN652611	.	.	.	-	-	-	-	-	-	-	-	T	.	.	T	.	C	.
EU827787	.	.	.	-	-	-	-	-	-	-	-	T	.	.	T	.	.	.

Genbank Accession Number	101	110	111	118	120	122	123	125	126	127	132	134	135	138
AY650768	A	G	T	T	A	A	C	T	G	T	T	A	T	C
MN652597	.	.	.	C	.	.	.	.	.	.	.	.	.	.
AY650778	.	.	.	C	.	.	.	.	.	.	.	.	.	.
AY650860	.	.	.	C	.	.	.	.	.	.	.	.	.	.
MN652598	.	.	.	.	.	.	.	.	.	.	.	.	.	.
MN652599	.	.	.	.	.	.	.	.	.	.	.	.	.	.
EU827759	.	.	.	.	.	.	.	.	.	.	.	.	.	.
MN652600	.	.	.	.	.	.	.	.	.	.	.	.	.	.
MN652601	.	.	.	.	.	.	C	.	.	.	.	.	.	.
AY650836	.	.	.	.	.	.	.	.	.	.	.	.	.	.
AY650821	.	.	.	.	.	.	.	.	.	.	.	.	.	.
MN652602	.	.	.	.	.	.	.	.	.	.	.	.	.	.
AY650858	.	.	.	.	.	.	.	.	.	.	.	.	.	.
MN652603	.	.	.	.	.	.	.	.	.	.	.	.	.	.
MN652604	C	.	.	.	.	.	.	.	T	.	.	.	.	.
MN652605	.	.	.	.	.	.	.	.	T	.	.	.	.	.
MN652606	.	.	.	.	.	.	.	.	.	.	.	.	.	G
MN652607	.	.	.	.	.	.	.	.	.	.	.	.	.	.
MN652608	.	.	.	.	.	.	.	.	.	.	.	.	C	.
AY650781	.	.	.	.	.	.	.	.	.	.	.	.	C	.
MN652609	.	.	.	.	.	.	.	.	.	.	A	.	C	.
AY650763	.	.	.	.	.	.	T	.	.	.	A	.	C	.
MN652595	.	.	.	.	.	.	T	.	.	.	A	.	C	.
AY650855	.	.	.	.	.	.	T	.	.	.	A	.	C	.
MN652596	.	.	.	.	.	.	T	.	.	.	A	.	C	.
AY650861	.	.	.	.	.	.	T	.	.	.	A	.	C	.
AY650809	.	.	.	.	G	G	.	A	C	C	.	C	.	G
MN652610	.	.	.	.	G	G	.	A	C	C	.	C	.	G
AY650762	.	.	.	.	G	G	.	A	C	C	.	C	.	G
AY650814	.	.	.	.	G	G	.	A	C	C	.	C	.	G
AY650805	.	.	.	.	G	G	.	A	C	C	.	C	.	G
AY650761	C	.	.	.	G	G	.	A	C	C	.	C	.	G
AY650829	.	.	.	.	G	G	.	A	C	C	.	C	.	G
EU827771	.	.	.	.	G	G	.	A	.	C	.	C	.	G
MN652611	.	.	.	.	G	G	.	A	C	C	.	C	.	G
EU827787	.	.	.	.	G	G	.	A	C	C	.	C	.	G

Genbank Accession Number	158	159	160	161	165	166	172	173	178	180	186	187	190	194	217	257	262
AY650768	A	T	T	T	A	A	A	T	C	C	-	T	G	G	T	A	G
MN652597	T	.	.	.	.	.	.	.	.	.	.	.	.	.	.	.	.
AY650778	T	.	.	.	.	.	.	.	.	.	.	.	.	.	.	.	.
AY650860	T	G	.	.	.	.	.	.	.	.	.	.	.	.	.	.	.
MN652598	.	.	.	.	.	.	.	.	.	.	.	.	.	.	.	.	.
MN652599	.	.	.	.	.	.	G	.	.	.	.	.	.	.	.	.	.
EU827759	.	.	.	.	.	.	.	.	.	.	.	.	.	.	.	.	.
MN652600	.	G	.	.	.	.	.	.	.	.	.	.	.	.	.	.	.
MN652601	.	G	.	.	.	.	.	.	.	.	.	.	.	.	.	.	.
AY650836	.	G	.	.	.	.	.	.	.	.	.	.	.	.	.	.	.
AY650821	.	C	.	.	.	.	.	.	.	.	.	.	.	.	.	.	.
MN652602	.	.	C	.	.	.	.	.	.	.	.	.	.	.	.	.	.
AY650858	.	.	.	.	.	.	.	.	.	.	.	.	.	.	.	.	.
MN652603	.	.	.	.	.	.	G	.	.	.	.	.	.	.	.	.	.
MN652604	.	.	.	.	.	.	.	.	.	.	.	.	.	.	.	.	.
MN652605	.	.	.	.	.	.	.	.	.	.	.	.	.	.	.	.	.
MN652606	.	.	.	.	.	.	.	.	.	.	.	.	.	.	.	.	.
MN652607	.	.	.	.	.	.	.	.	.	.	.	.	.	.	.	.	.
MN652608	T	.	.	.	.	.	G	T	.	.	.	.	A	.	.	.	.
AY650781	T	.	.	.	.	.	G	T	.	.	.	.	A	.	.	.	.
MN652609	T	.	.	.	.	.	G	T	.	.	.	.	A	.	.	.	.
AY650763	T	.	.	.	.	C	G	T	.	.	.	.	A	.	.	.	A
MN652595	T	.	.	.	.	C	G	T	.	.	.	.	A	.	.	C	A
AY650855	T	.	.	.	.	C	G	T	.	.	.	.	A	C	.	.	A
MN652596	T	.	.	.	.	C	G	T	.	.	.	.	A	.	.	.	A
AY650861	T	.	.	.	G	C	G	T	.	.	.	.	A	.	.	.	A
AY650809	T	G	C	.	.	.	.	.	.	.	T	.	.	.	A	G	A
MN652610	T	G	C	.	.	.	.	.	.	.	T	.	.	.	A	G	A
AY650762	T	G	C	.	C	.	.	.	.	.	T	.	.	.	A	G	A
AY650814	T	G	C	C	.	.	.	.	.	.	T	.	.	.	A	G	A
AY650805	T	G	C	.	.	G	.	.	.	.	T	.	.	.	A	G	A
AY650761	T	G	C	.	.	.	.	.	.	.	T	.	.	.	A	G	A
AY650829	T	G	C	.	.	.	.	.	.	.	T	.	.	.	A	G	A
EU827771	T	G	C	C	.	.	.	.	.	.	T	.	.	.	A	G	A
MN652611	T	G	C	.	.	.	.	.	.	.	T	.	.	.	A	G	A
EU827787	T	G	C	.	.	.	.	.	.	.	T	.	.	.	A	G	.

According to Bremer et al. [12] the mtDNA haplotypes of swordfish clustered into two highly divergent clades, Clade I and Clade II. The MJ network (Figure 2) identified the two mtDNA clades.

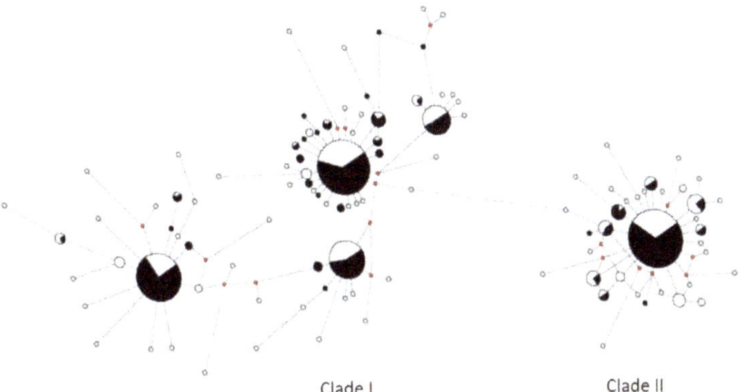

Figure 2. MJ network for the complete dataset of Mediterranean swordfish. White slices indicate haplotype observed by Viñas [17] and black slices recent sample (this work). Small red circles correspond to missing (or hypothetical) haplotypes.

Clade I included four centroids (Genbank n. AY650768, AY650763, AY650778, AY650781) represented respectively by 61, 55, 14 and 25 fish, while Clade II was characterized by a single star-like formation featuring one major centroid (AY650762) with a total of 73 individuals. These five centroids haplotypes were observed with high frequencies in all Mediterranean localities investigated in this study, as already observed in previous works [10,17]. The other haplotypes, instead, were represented in very few numbers of the individuals often representing singletons. In this study seventeen, previously unidentified haplotypes were sampled. Fifteen of which belong to Clade I and two belong to Clade II and all were observed at low frequencies (Table S1). Sequences are available from the NCBI database with accession numbers MN652595-MN652611.

Comparison among historical dataset from Viñas et al. [17] and the present dataset, respectively composed by 251 and 287 swordfish, revealed the same percentage of individuals belonging to Clade I (about 66%) and Clade II (about 33%). However, dataset comparison indicates a reduction in the number of haplotypes from 93 in the historical to 36 in the contemporary samples (Table 3).

Table 3. Summary of molecular diversity indices for Mediterranean swordfish temporal samples for pooled clades and each clade separately. The number of individuals (N) and haplotypes (H), the haplotype diversity (h), the nucleotide diversity (π) and effective population size parameter (θ) are shown.

Title	N	H	h (s.d.)	π (s.d.)	θ (95% C.I.)
Viñas et al., 2010					
Clade I + II	251	93	0.946 (0.007)	0.160 (0.080)	0.142 (0.121–0.188)
Clade I	156	60	0.910 (0.014)	0.034 (0.020)	0.076 (0.060–0.107)
Clade II	95	33	0.861 (0.031)	0.006 (0.004)	0.044 (0.0.29–0.062)
This study					
Clade I + II	286	36	0.844 (0.010)	0.130 (0.070)	0.033 (0.028–0.044)
Clade I	190	26	0.793 (0.018)	0.029 (0.015)	0.019 (0.015–0.030)
Clade II	96	10	0.418 (0.063)	0.003 (0.002)	0.010 (0.006–0.017)

Haplotype reduction was confirmed by the rarefaction curve that reported the highest haplotype richness values in the older sample (Figure 3).

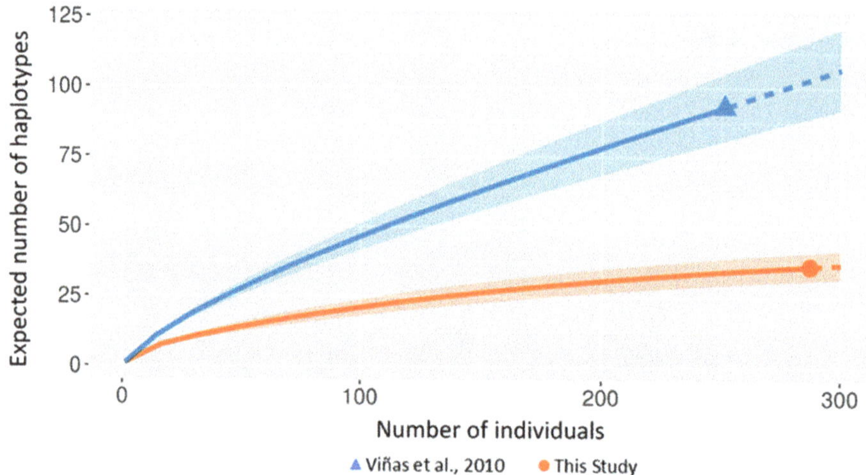

Figure 3. Size-based rarefaction (solid curves) and extrapolation (dashed curves) with 95% confidence intervals (shaded areas) comparing ancient Mediterranean swordfish (blue) and contemporary (orange) samples.

Considering the two clades separately, the observed haplotype number decreased from 60 to 26 for Clade I and from 33 to 10 for Clade II. Nineteen haplotypes were shared between datasets (11 for Clade I and 8 for Clade II). Moreover, in the modern sample, a shift towards the fixation of the main five haplotypes (i.e., centroids) was also observed (Figure 2, Table S2).

A reduction of haplotype diversity was observed between historical ($h = 0.947 \pm 0.007$ s.d.) and modern datasets ($h = 0.844 \pm 0.01$ s.d.). Haplotype diversity decrease was more evident considering the two clades separately. For Clade I the diversity in historical samples ($h = 0.91$) was greater than the contemporary ($h = 0.793$) as well as for Clade II where diversity was $h = 0.861$ in the historical sample and $h = 0.418$ in the modern sample. Levels of nucleotide diversity were similar throughout the two databases. All estimates of haplotype and nucleotide diversity are given in Table 3. The AMOVA analysis identifies significant genetic differentiation between older and recent datasets. Low levels of differentiation ($F_{ST} = 0.018$, $p < 0.001$) were detected between samples pooling clades and considering Clade I singularly, while historical and modern samples were moderately differentiated ($F_{ST} = 0.121$, $p < 0.001$) considering only Clade II (Table 4).

Table 4. AMOVA results of Mediterranean swordfish temporal sample comparison.

Source of Variation	Variance Component	Percentage of Variation	Fixation Index (ϕ_{ST})	*p*-Value
Whole dataset				
Among temporal samples	0.008	1.77	0.0177	0.0000
Within samples	0.446	98.23		
Clade I				
Among temporal samples	0.008	1.85	0.0184	0.0005
Within samples	0.423	98.15		
Clade II				
Among temporal samples	0.044	12.10	0.121	0.0000
Within samples	0.319	87.90		

The older sample significantly differed from the modern one when compared through an exact test of sample differentiation. The effective population size parameters of swordfish populations were estimated for each mtDNA CR-I clade separately. The comparison of the θ values showed a reduction of the relative females' effective population size in the recent sample (θ = 0.033, 95% C.I. 0.028–0.044) compared to the historical one (θ = 0.142, 95% C.I. 0.121–0.188) (Table 3). Ratios between the historical and current θ suggested a reduction of approximately three-quarters of the female population size. Estimates were congruent considering both the clades separately and pooled them together.

4. Discussion

In this study, for the first time, a temporal approach has been used to study recent changes in mitochondrial genetic diversity in the threatened Mediterranean swordfish. Comparison between temporal samples allows us to identify short-term changes that happened in the last few decades. Our results revealed that Mediterranean swordfish has undergone genetic depletion during the last twenty years.

The number of mitochondrial haplotypes observed in the recent sample (36 in total; 26 Clade I and 10 Clade II) was largely reduced than in previous studies [10,17]. The PCR-SSCP is a valuable tool to detect nucleotidic variation in DNA because even a single nucleotide change may alter the conformation of ssDNA. Despite this high resolving power, there is evidence that not all nucleotide modifications result in conformation changes [38]. To exclude loss of haplotypes due to the SSCP analysis we sequenced all the individuals belonging to the Clade II and all sequences confirmed the SSCP interpretation. Therefore, we consider the SSCP technique used in this paper sensitive enough to detect each single nucleotide polymorphism.

The temporal comparison based on the mtDNA genetic variability highlighted as approximately one-third of the haplotypes was lost in the last twenty years. This reduction was noticeable as a lower haplotype diversity observed on modern samples (h = 0.84, this study) compared to previous ones (h = 0.94) [10,17]. When the two mitochondrial clades were analysed separately, the reduction of genetic diversity was not uniformly distributed between clades and the reduction in genetic diversity was more evident for the Clade II. Diversity comparison indicates approximately half less haplotype diversity in the contemporary Clade II samples than in the historical. Genetic differentiation between contemporary and historical swordfish was corroborated by both AMOVA and exact test results (Table 4).

The reduction in haplotype diversity with an increment in the proportion of identical haplotype and the loss of rare haplotypes observed in current Mediterranean swordfish suggest that population has undergone a demographic reduction [28]. As far as it's known, nowadays, the Mediterranean swordfish stock is considered threatened by overfishing [7]. In fact, the spawning stock biomass (SSB) estimated is less than 15% of the maximum sustainable yield (B_{MSY}). From 2007–2010 the mesopelagic longline was introduced in Mediterranean swordfish fishery, substituting the traditional surface longline [7]. The two types of gears differ mainly for the depth of displacement, with the mesopelagic reaching deeper depths (50–800 m) than the traditional longline (15–60 m). Increment in fish size catch using the mesopelagic longline has revealed the presence of a bulk of large spawners, that found their refuge in the deep [39]. However, after only six years the mean swordfish size exploited shifted towards smaller sizes, in a situation very similar to that recorded in the past for surface driftnets and surface longline [40]. Thus, it is possible to assume that the mesopelagic gear catching bigger swordfish may have increased the impact on adults, affecting the spawning stock [39–41]. Specifically, mtDNA is maternally inherited therefore, the analysis of mtDNA variability reflects only the history of the Mediterranean swordfish female portion. Swordfish females reach larger sizes than males, outnumbering males in larger size classes (LJFL > 140 cm) [5,39,42]. Therefore, an increase in the average size may be related to an increase in the number of mature females exploited. Furthermore, females mature at larger age and size than males, about two years later. The high percentage (50–70%) of small fish, often still immature, reported by the annual catch estimates, may affect the number of recruited new females in the spawning stock. Comparison of θ_s values supports the scenario of a recent population size decrease in Mediterranean swordfish stock, suggesting a reduction of approximately three-quarters of effective female population size (Table 3). Thus, the fishery may have altered the sex ratio, reducing the number of females in the breeding population, therefore leading to the loss in mitochondrial diversity.

Overexploitation and environmental degradation were identified as the main causes of reductions in stocks and extinction of marine species [43,44]. Human activities may reduce the genetic variability of the population in an extremely short period [45]. Overexploitation has been related to a decline in genetic diversity across a wide range of marine fish species [22]. The genetic diversity estimated in harvested and unharvested fish stocks over 140 species, resulted to be lower in the exploited than in the non-harvest stocks [22]. However, a limited number of studies have assessed the temporal reduction of effective population size (N_e) and the loss of genetic variation related to harvesting due to lack of historical data as well as samples. Some of these studies, using DNA extracted from archived otoliths or scales, found a loss of neutral genetic diversity in some fish species. A significant reduction in genetic diversity was described in New Zealand *Hoplostethus atlanticus* over only 6 years, during which time the exploitation reduced the biomass by 70% [46]. Marked genetic changes were also detected using microsatellite in the *Gadus morhua* Flamborough Head population across a period during which the population exhibited a decline in SSB related to high levels of exploitation [27]. Hauser [25] detected a significant decline in genetic diversity in a New Zealand snapper population over the 50 years since the onset of exploitation. Moreover, in Adriatic European anchovy (*Engraulis encrasicolus*) a significant reduction of mean genetic parameters have been detected and related to the stock collapse in 1987 [47]. The number of fish in a population (census population size, N) is often much larger than the fish that are reproducing and helping to maintain genetic diversity (the genetically effective population size, N_e). Hauser [25] using the N_e/N estimate in snapper suggested that fish stocks of several million individuals may be in danger of losing genetic variability in the long term. Considering that, globally 31% of fish populations are fished unsustainably, with an additional 60% fully fished [48], fishing may have already caused a considerable loss of overall biodiversity.

This study reports the first direct measure of reduction in genetic diversity for Mediterranean swordfish during a short period, as measured both in the direct loss of mitochondrial haplotypes and reductions in haplotype diversity. Genetic changes observed in this study suggest that reduction in the SSB of Mediterranean swordfish caused significant changes in genetic composition. The significant loss of mitochondrial diversity in Mediterranean swordfish over a short period is alarming because the rapid loss of genetic diversity has been shown to have harmful effects. Reduction in genetic diversity, coupled with low population size [20] and reproductive isolation [2], can disfavour Mediterranean swordfish recovery. This may result in genetic drift and inbreeding depression and reduction in fitness limiting long-term adaptability [23,24], particularly if abundance remains low and diversity continues to decay [22]. This result underlines the urgent necessity to re-evaluate management strategies of the swordfish in the Mediterranean Sea.

Supplementary Materials: The following are available online at http://www.mdpi.com/1424-2818/12/5/170/s1, Table S1. Haplotype frequencies observed have been reported for each sample locality, Table S2. GenBank accession numbers of the mtDNA CR haplotypes, clade assignment and frequencies observed in each temporal sample, Table S3. Predicted fragment sizes (bp) of a 290 bps long portion of the mitochondrial control region virtually double-digested with VspI and PacI, Table S4. Variable sites in the control-region segment of all 110 haplotypes of swordfish mtDNA detected in Viñas et al., 2010 and this work (in bold). Numbers above sites refer to nucleotide position on the L-strand of the control region of the swordfish. Identity with sequence 1 is indicated by dots and deletions of 1 residue with dashes.

Author Contributions: Conceptualization, T.R.; Methodology, T.R.; Formal analysis, T.R.; Investigation, T.R.; Resources, G.G. and V.C.B.; Data curation, E.C., T.R.; Writing—Original draft preparation, T.R.; Writing—Review and editing, A.S., T.F. and T.R.; Visualization, T.R.; Supervision, V.C.B.; Project administration, G.G.; Funding acquisition, O.C. All authors have read and agreed to the published version of the manuscript.

Funding: This work was funded by the Ministry of Agriculture, Food and Forestry Policies (MIPAAF), note 6775, Art.36 Paragraph 1 Reg (UE9 n 508/2014) to O.C.

Acknowledgments: We truly thank the fishermen, Pesca Pronta Import-Export s.r.l. for their extremely precious collaboration during sampling activities. The authors wish to thank all the staff at OCEANIS s.r.l. (Ercolano, Naples) for the support in sampling activities.

Conflicts of Interest: The authors declare no conflict of interest.

References

1. Palko, B.J.; Beardsley, G.L.; Richards, W.J. Synopsis of the biology of the swordfish, *Xiphias gladius* (L.). U.S. Dep. Commer. *NOAA Tech. Rep. NMFS Circ.* **1981**, *441*, 21.
2. Neilson, J.; Arocha, F.; Cass-Calay, S.; Mejuto, J.; Ortiz, M.; Scott, G.; Smith, C.; Travassos, P.; Tserpes, G.; Andrushchenko, I. The recovery of Atlantic swordfish: The comparative roles of the regional fisheries management organization and species biology. *Rev. Fish. Sci.* **2013**, *21*, 59–97. [CrossRef]
3. Abascal, F.J.; Mejuto, J.; Quintans, M.; García-Cortés, B.; Ramos-Cartelle, A. Tracking of the broadbill swordfish, *Xiphias gladius*, in the central and eastern North Atlantic. *Fish. Res.* **2015**, *162*, 20–28. [CrossRef]
4. Arocha, F. Swordfish Reproduction in the Atlantic Ocean: An Overview. *GCR* **2007**, 19. [CrossRef]
5. Marisaldi, L.; Basili, D.; Candelma, M.; Sesani, V.; Pignalosa, P.; Gioacchini, G.; Carnevali, O. Maturity assignment based on histology-validated macroscopic criteria: Tackling the stock decline of the Mediterranean swordfish (*Xiphias gladius*). *Aquat. Conserv. Mar. Freshw. Ecosyst.* **2020**, *30*, 303–314. [CrossRef]
6. Macías, D.; Hattour, A.; De la Serna, J.M.; Gómez-Vives, M.J.; Godoy, D. Reproductive characteristics of swordfish (*Xiphias gladius*) caught in the southwestern Mediterranean during 2003. *Collect. Vol. Sci. Pap. ICCAT* **2005**, *58*, 454–469.
7. Anon. *Report for Biennial Period, 2018–19 PART I (2018)—Volume 2*; International Commission for the Conservation of Atlantic Tunas: Madrid, Spain, 2019; pp. 182–192.
8. Bremer, J.A.; Mejuto, J.; Gómez-Márquez, J.; Pla-Zanuy, C.; Viñas, J.; Marques, C.; Hazin, F.; Griffiths, M.; Ely, B.; García-Cortés, B. Genetic population structure of Atlantic swordfish: Current status and future directions. *Collect. Vol. Sci. Pap. ICCAT* **2007**, *61*, 107–118.
9. Viñas, J.; Bremer, J.A.; Pla, C. Inter-oceanic genetic differentiation among albacore (*Thunnus alalunga*) populations. *Mar. Biol.* **2004**, *145*, 225–232. [CrossRef]
10. Bremer, J.R.A.; Viñas, J.; Mejuto, J.; Ely, B.; Pla, C. Comparative phylogeography of Atlantic bluefin tuna and swordfish: The combined effects of vicariance, secondary contact, introgression, and population expansion on the regional phylogenies of two highly migratory pelagic fishes. *Mol. Phylogenetics Evol.* **2005**, *36*, 169–187. [CrossRef]
11. Šegvić-Bubić, T.; Marrone, F.; Grubišić, L.; Izquierdo-Gomez, D.; Katavić, I.; Arculeo, M.; Brutto, S.L. Two seas, two lineages: How genetic diversity is structured in Atlantic and Mediterranean greater amberjack *Seriola dumerili* Risso, 1810 (Perciformes, Carangidae). *Fish. Res.* **2016**, *179*, 271–279. [CrossRef]
12. Bremer, J.R.A.; Mejuto, J.; Greig, T.W.; Ely, B. Global population structure of the swordfish (*Xiphias gladius* L.) as revealed by analysis of the mitochondrial DNA control region. *J. Exp. Mar. Biol. Ecol.* **1996**, *197*, 295–310. [CrossRef]
13. Rosel, P.E.; Block, B.A. Mitochondrial control region variability and global population structure in the swordfish, *Xiphias gladius*. *Mar. Biol.* **1996**, *125*, 11–22. [CrossRef]
14. Chow, S.; Okamoto, H.; Uozumi, Y.; Takeuchi, Y.; Takeyama, H. Genetic stock structure of the swordfish (*Xiphias gladius*) inferred by PCR-RFLP analysis of the mitochondrial DNA control region. *Mar. Biol.* **1997**, *127*, 359–367. [CrossRef]
15. Muths, D.; Grewe, P.; Jean, C.; Bourjea, J. Genetic population structure of the Swordfish (*Xiphias gladius*) in the southwest Indian Ocean: Sex-biased differentiation, congruency between markers and its incidence in a way of stock assessment. *Fish. Res.* **2009**, *97*, 263–269. [CrossRef]
16. Reeb, C.A.; Arcangeli, L.; Block, B.A. Structure and migration corridors in Pacific populations of the Swordfish *Xiphius gladius*, as inferred through analyses of mitochondrial DNA. *Mar. Biol.* **2000**, *136*, 1123–1131. [CrossRef]
17. Viñas, J.; Pérez-Serra, A.; Vidal, O.; Alvarado Bremer, J.R.; Pla, C. Genetic differentiation between eastern and western Mediterranean swordfish revealed by phylogeographic analysis of the mitochondrial DNA control region. *ICES J. Mar. Sci.* **2010**, *67*, 1222–1229. [CrossRef]
18. Bremer, J.R.A.; Baker, A.J.; Mejuto, J. Mitochondrial DNA control region sequences indicate extensive mixing of swordfish (*Xiphias gladius*) populations in the Atlantic Ocean. *Can. J. Fish. Aquat. Sci.* **1995**, *52*, 1720–1732. [CrossRef]

19. Pujolar, J.M.; Roldan, M.I.; Pla, C. A genetic assessment of the population structure of swordfish (*Xiphias gladius*) in the Mediterranean Sea. *J. Exp. Mar. Biol. Ecol.* **2002**, *276*, 19–29. [CrossRef]

20. Kotoulas, G.; Mejuto, J.; Antoniou, A.; Kasapidis, P.; Tserpes, G.; Piccinetti, C.; Peristeraki, P.; Garcia-Cortes, B.; Oikonomaki, K.; De la Serna, J.M. Global genetic structure of swordfish (*Xiphias gladius*) as revealed by microsatellite DNA markers. *Collect. Vol. Sci. Pap. ICCAT* **2007**, *61*, 79–88.

21. Xiphias gladius (errata version published in 2016). The IUCN Red List of Threatened Species 2011: e.T23148A88828055. Available online: https://dx.doi.org/10.2305/IUCN.UK.2011-2.RLTS.T23148A9422329.en. (accessed on 17 December 2017).

22. Pinsky, M.L.; Palumbi, S.R. Meta-analysis reveals lower genetic diversity in overfished populations. *Mol. Ecol.* **2014**, *23*, 29–39. [CrossRef]

23. Frankham, R. Genetics and extinction. *Biol. Conserv.* **2005**, *126*, 131–140. [CrossRef]

24. Spielman, D.; Brook, B.W.; Frankham, R. Most species are not driven to extinction before genetic factors impact them. *Proc. Natl. Acad. Sci. USA* **2004**, *101*, 15261–15264. [CrossRef]

25. Hauser, L.; Adcock, G.J.; Smith, P.J.; Ramírez, J.H.B.; Carvalho, G.R. Loss of microsatellite diversity and low effective population size in an overexploited population of New Zealand snapper (*Pagrus auratus*). *Proc. Natl. Acad. Sci. USA* **2002**, *99*, 11742–11747. [CrossRef]

26. Turner, T.F.; Wares, J.P.; Gold, J.R. Genetic effective size is three orders of magnitude smaller than adult census size in an abundant, estuarine-dependent marine fish (*Sciaenops ocellatus*). *Genetics* **2002**, *162*, 1329–1339.

27. Hutchinson, W.F.; Oosterhout, C.; Rogers, S.I.; Carvalho, G.R. Temporal analysis of archived samples indicates marked genetic changes in declining North Sea cod (*Gadus morhua*). *Proc. R. Soc. Lond. Ser. B Biol. Sci.* **2003**, *270*, 2125–2132. [CrossRef]

28. Allendorf, F.W.; England, P.R.; Luikart, G.; Ritchie, P.A.; Ryman, N. Genetic effects of harvest on wild animal populations. *Trends Ecol. Evol.* **2008**, *23*, 327–337. [CrossRef] [PubMed]

29. Thompson, J.D.; Gibson, T.J.; Higgins, D.G. Multiple sequence alignment using ClustalW and ClustalX. *Curr. Protoc. Bioinform.* **2003**, *1*, 2–3. [CrossRef] [PubMed]

30. Hall, T.A. BioEdit: A user-friendly biological sequence alignment editor and analysis program for Windows 95/98/NT. *Nucleic Acids Symp. Ser.* **1999**, *41*, 95–98.

31. Excoffier, L.; Lischer, H.E. Arlequin suite ver 3.5: A new series of programs to perform population genetics analyses under Linux and Windows. *Mol. Ecol. Resour.* **2010**, *10*, 565–567. [CrossRef] [PubMed]

32. Heck, K.L., Jr.; van Belle, G.; Simberloff, D. Explicit calculation of the rarefaction diversity measurement and the determination of sufficient sample size. *Ecology* **1975**, *56*, 1459–1461. [CrossRef]

33. Hsieh, T.C.; Ma, K.H.; Chao, A. iNEXT: An R package for rarefaction and extrapolation of species diversity (H ill numbers). *Methods Ecol. Evol.* **2016**, *7*, 1451–1456. [CrossRef]

34. Chao, A.; Gotelli, N.J.; Hsieh, T.C.; Sander, E.L.; Ma, K.H.; Colwell, R.K.; Ellison, A.M. Rarefaction and extrapolation with Hill numbers: A framework for sampling and estimation in species diversity studies. *Ecol. Monogr.* **2014**, *84*, 45–67. [CrossRef]

35. Bandelt, H.-J.; Forster, P.; Röhl, A. Median-joining networks for inferring intraspecific phylogenies. *Mol. Biol. Evol.* **1999**, *16*, 37–48. [CrossRef] [PubMed]

36. Excoffier, L.; Smouse, P.E.; Quattro, J.M. Analysis of molecular variance inferred from metric distances among DNA haplotypes: Application to human mitochondrial DNA restriction data. *Genetics* **1992**, *131*, 479–491. [PubMed]

37. Kuhner, M.K. LAMARC 2.0: Maximum likelihood and Bayesian estimation of population parameters. *Bioinformatics* **2006**, *22*, 768–770. [CrossRef]

38. Hayashi, K.; Yandell, D.W. How sensitive is PCR-SSCP? *Hum. Mutat.* **1993**, *2*, 338–346. [CrossRef]

39. Garibaldi, F. Effects of the introduction of the mesopelagic longline on catches and size structure of swordfish in the Ligurian sea (western Mediterranean). *Collect. Vol. Sci. Pap. ICCAT* **2015**, *71*, 2006–2014.

40. Garibaldi, F.; Lanteri, L. Notes about a tagged/recaptured swordfish in the Ligurian Sea (western Mediterranean). *Collect. Vol. Sci. Pap. ICCAT* **2017**, *74*, 1354–1361.

41. Bertolino, F.; Camolese, C.; Dell'Aquila, M.; Mariani, A.; Valastro, M. Swordfish (*Xiphias gladius* L.) fisheries using mesopelagic longline in the Mediterranean Sea by Italian fishing fleet. *Collect. Vol. Sci. Pap. ICCAT* **2015**, *71*, 2073–2078.

42. Abid, N.; Laglaoui, A.; Arakrak, A.; Bakkali, M. The reproductive biology of swordfish (*Xiphias gladius*) in the Strait of Gibraltar. *J. Mar. Biol. Assoc. U. K.* **2019**, *99*, 649–659. [CrossRef]

43. Dulvy, N.K.; Sadovy, Y.; Reynolds, J.D. Extinction vulnerability in marine populations. *Fish Fish.* **2003**, *4*, 25–64. [CrossRef]
44. Dulvy, N.K.; Jennings, S.; Rogers, S.I.; Maxwell, D.L. Threat and decline in fishes: An indicator of marine biodiversity. *Can. J. Fish. Aquat. Sci.* **2006**, *63*, 1267–1275. [CrossRef]
45. Kenchington, E.L. 14 The Effects of Fishing on Species and Genetic Diversity. *Responsible Fish. Mar. Ecosyst.* **2003**, 235.
46. Smith, P.J.; Francis, R.; McVeagh, M. Loss of genetic diversity due to fishing pressure. *Fish. Res.* **1991**, *10*, 309–316. [CrossRef]
47. Ruggeri, P.; Splendiani, A.; Di Muri, C.; Fioravanti, T.; Santojanni, A.; Leonori, I.; De Felice, A.; Biagiotti, I.; Carpi, P.; Arneri, E. Coupling demographic and genetic variability from archived collections of European anchovy (*Engraulis encrasicolus*). *PLoS ONE* **2016**, *11*, e0151507. [CrossRef] [PubMed]
48. FAO. *The state of world fisheries and aquaculture 2016. Contributing to Food Security and Nutrition for all*; FAO: Rome, Italy, 2016.

MDPI
St. Alban-Anlage 66
4052 Basel
Switzerland
Tel. +41 61 683 77 34
Fax +41 61 302 89 18
www.mdpi.com

Diversity Editorial Office
E-mail: diversity@mdpi.com
www.mdpi.com/journal/diversity